DIGITAL ELECTRONICS

For the Hobbyist, Technician & Engineer

Asheville-Buncombe
Technical Community College
Learning Resources Center
340 Victoria Road
Asheville, NC 28801

DIGITAL ELECTRONICS

For the Hobbyist, Technician & Engineer

By Stephen Kamichik
B.Sc., E.E.T., B.Eng., M.Eng.

An Imprint of
Howard W. Sams & Company
Indianapolis, Indiana

®1996 by Stephen Kamichik

FIRST EDITION—1996

PROMPT® Publications is an imprint of Howard W. Sams & Company, a Bell Atlantic company, 2647 Waterfront Parkway, East Drive, Suite 300, Indianapolis, IN 46214-2041.

All rights reserved. No part of this book shall be reproduced, stored in a retrieval system, or transmitted by any means, electronic, mechanical, photocopying, recording, or otherwise, without written permission from the publisher. No patent liability is assumed with respect to the use of the information contained herein. While every precaution has been taken in the preparation of this book, the author, the publisher, or the seller assumes no responsibility for errors or omissions. Neither is any liability assumed for damages resulting from the use of information contained herein.

International Standard Book Number: 0-7906-1075-2

Library of Congress Catalog Card Number: 96-67470

Acquisitions Editor: Candace M. Drake
Editor: Natalie F. Houck
Assistant Editors: Karen Mittelstadt, Pat Brady
Illustrators: Natalie Houck, Christy Pierce, Terry Varvel
Typesetter: Leah Marckel
Cover Design: Kelli Ternet

Trademark Acknowledgments:
All terms mentioned in this book that are known or suspected to be trademarks or services have been appropriately capitalized. PROMPT® Publications, Howard W. Sams & Company, and Bell Atlantic cannot attest to the accuracy of this information. Use of a term in this book should not be regarded as affecting the validity of any trademark or service mark.

Printed in the United States of America

2 3 4 5 6 7 8 9

TABLE OF CONTENTS

INTRODUCTION 1

Chapter One
ANALOG vs. DIGITAL 3
    Example 1-1 5
  PROBLEMS 5

Chapter Two
LOGIC GATES 7
  INVERTER (NOT GATE) 7
  OR GATE 7
  AND GATE 8
  XOR GATE 8
  PROBLEMS 8

Chapter Three
LOGIC FAMILIES 11
  RESISTOR TRANSISTOR LOGIC (RTL) 13
  DIODE TRANSISTOR LOGIC (DTL) 13
  TRANSISTOR-TRANSISTOR LOGIC (TTL) 14
  EMITTER COUPLED LOGIC (ECL) 15
  COMPLEMENTARY METAL-OXIDE
      SEMICONDUCTOR LOGIC (CMOS) 16
  PROBLEMS 16

Chapter Four
LOGIC FUNCTION IMPLEMENTATION 19
  SUM OF PRODUCTS 20
  PRODUCT OF SUMS 20
  WIRED LOGIC 20
    Example 4-1 20

KARNAUGH MAP	21
Example 4-2	22
Example 4-3	22
Example 4-4	24
ANALYZING MULTILEVEL CIRCUITS	24
Example 4-5	24
Example 4-6	24
Example 4-7	26
PROBLEMS	26

Chapter Five
FLIP-FLOPS 29

FLIP-FLOP	30
CLOCKED AND GATED FLIP-FLOP	30
MASTER-SLAVE FLIP-FLOP	33
DIRECT CLEAR AND DIRECT SET INPUTS	33
D FLIP-FLOP	34
JK FLIP-FLOP	34
PROBLEMS	34

Chapter Six
CONTROL CIRCUITS 37

Example 6-1	37
Example 6-2	39
PROBLEMS	40

Chapter Seven
CODES 41

Example 7-1	41
Example 7-2	41
OCTAL NUMBER SYSTEM	42
Example 7-3	42
BINARY ARITHMETIC	42
Example 7-4	42
Example 7-5	42
Example 7-6	43
BINARY CODES	43

8421 CODES 43
    Example 7-7 43
    Example 7-8 44
BCD CODE 44
EXCESS-3 CODE 45
2421 CODE 45
GRAY CODE 45
    Example 7-9 45
    Example 7-10 46
ERROR DETECTION AND CORRECTION 46
BIQUINARY CODE 46
ASCII CODE 46
HAMMING CODE 46
PROBLEMS 47

Chapter Eight
REGISTERS 49
PARALLEL ENTRY 49
TIMING 50
    Example 8-1 50
    Example 8-2 50
    Example 8-3 50
    Example 8-4 50
    Example 8-5 53
SERIAL ENTRY 53
COUNTERS 53
RING COUNTER 54
COMPLEMENTING RING COUNTER 54
DIMINISHED COMPLEMENTING RING COUNTER .. 55
PROBLEMS 55

Chapter Nine
ENCODERS, DECODERS & MULTIPLEXERS 57
ENCODERS 57
DECODERS 57
    Rectangular Decoder 58
    Tree Decoder 58

Dual Tree Decoder	*59*
1/n Decoder	*60*
1/2 Decoder	*60*
1/n² Decoder	*60*
1/10 Decoder	*60*
MULTIPLEXERS	*63*
APPLICATIONS FOR MULTIPLEXERS	*63*
DEMULTIPLEXERS	*64*
PROBLEMS	*66*

Chapter Ten
COMPARATOR & EXCLUSIVE-OR CIRCUITS *69*
    APPLICATIONS *71*
    COMPARATOR CIRCUITS *73*
    PROBLEMS *73*

Chapter Eleven
COUNTERS *75*
    RIPPLE COUNTERS *75*
    SYNCHRONOUS COUNTERS *77*
    COUNTER DESIGN *80*
        Example 11-1 .... *80*
        Example 11-2 .... *80*
    PROBLEMS *84*

Chapter Twelve
ARITHMETIC CIRCUITS *85*
    RIPPLE ADDER *85*
    LOOK AHEAD ADDER *87*
    APPLICATIONS *89*
    PROBLEMS *91*

Chapter Thirteen
MEMORY *93*
    MEMORY EXPANSION *96*
    PROBLEMS *97*

Chapter Fourteen
DIGITAL-TO-ANALOG
& ANALOG-TO-DIGITAL CONVERTERS 101
DIGITAL-TO-ANALOG CONVERTERS *101*
Binary Weighted Ladder D/A Converter *103*
R-2R Ladder D/A Converter *103*
ANALOG-TO-DIGITAL CONVERTERS *103*
Parallel A/D Converter *103*
Successive Approximation A/D Converter *104*
Dual Slope A/D Converter *104*
Counter A/D Converter *106*
PROBLEMS *107*

Chapter Fifteen
THE FUTURE OF DIGITAL ELECTRONICS 109

Appendix
PROBLEM SOLUTIONS 111

GLOSSARY 141

INDEX 147

ABOUT THE AUTHOR

Stephen Kamichik received his Electrical Engineering Technology diploma from Ryerson Polytechnical Institute in 1975. He received his Bachelor and Master of Engineering from Concordia University in 1986 and 1989, respectively. Mr. Kamichik worked for several years as a professional technician, including a four-year stint at SPAR, where he was involved with prototyping the original Canadarm. He also worked as an engineer designing custom electronic circuits. Mr. Kamichik holds Canadian and American patents for a coded power control system. He has designed everything from home theater systems to test equipment. His first book, *Advanced Electronic Projects for Your Home & Automobile*, details the construction of some of his original circuits.

Mr. Kamichik has published two other books with PROMPT Publications in addition to this one: *Advanced Electronic Projects for Your Home & Automobile* and *Semiconductor Essentials for Hobbyists, Technicians & Engineers*.

Mr. Kamichik has several hobbies. He enjoys bowling, and has an average of about 176. He has a twenty-year-old Datsun 280Z that he maintains himself, and has won several trophies at car shows. He has formed a Datsun sports car club in his region in Quebec, Canada. Mr. Kamichik and his wife, Gail, have recently designed and built their dream home. The house was designed as a passive solar home to take advantage of free solar energy to heat their home.

INTRODUCTION

When I first studied digital electronics, there were no books recommended by the schools I attended. This text can be used to supplement the lectures of an introductory course to digital electronics.

This book was written as a textbook for a first course in digital electronics. It can also serve as a review for practicing technicians and engineers. The hobbyist can also learn about digital electronics because no prior knowledge is assumed in this text.

Chapter 1 is an introduction to digital electronics and the difference between digital and analog circuits. Chapter 2 discusses logic gates. Chapter 3 deals with logic families. Chapter 4 explains logic function implementation. Chapter 5 describes flip-flops.

Chapter 6 is about control circuits. Chapter 7 deals with codes. Chapter 8 is about registers. Chapter 9 describes encoders, decoders and multiplexers. Chapter 10 discusses comparators and exclusive-OR circuits. Chapter 11 deals with counters. Chapter 12 is about arithmetic circuits. Chapter 13 describes memory and chapter 14 discusses digital-to-analog and analog-to-digital converters. Chapter 15 discusses the possible future of digital electronics.

Each chapter is a lesson in digital electronics, with a problem set at the end of the chapter to test the reader's understanding of the material presented.

If the reader has the equipment, he or she can build the circuits described in this book to verify their operation. Building and testing a circuit is the best way to fully understand its operation.

Chapter One
ANALOG vs. DIGITAL

The universe obeys physical laws which are mostly *analog*. Acceleration and deceleration are continuous processes. The planets orbit around the sun in continuous elliptical orbits. Biological functions such as breathing are continuous and therefore analog in nature.

Many devices in our everyday world operate in a *digital* manner. The points of the ignition of an automobile are either OPENED or CLOSED, and therefore operate digitally.

An *analog signal* has an infinite number of levels. Therefore, a *sinusoidal waveform* is an analog signal because it has an infinite number of levels. *Figure 1-1* shows a sinusoidal waveform. The reader should note that the sinusoidal waveform has an infinite number of levels including its minimum and maximum values.

A *square waveform* is a two-level signal because it changes from its minimum value to its maximum value instantaneously. A two-level signal is also known as a *digital signal*. *Figure 1-2* shows a square waveform.

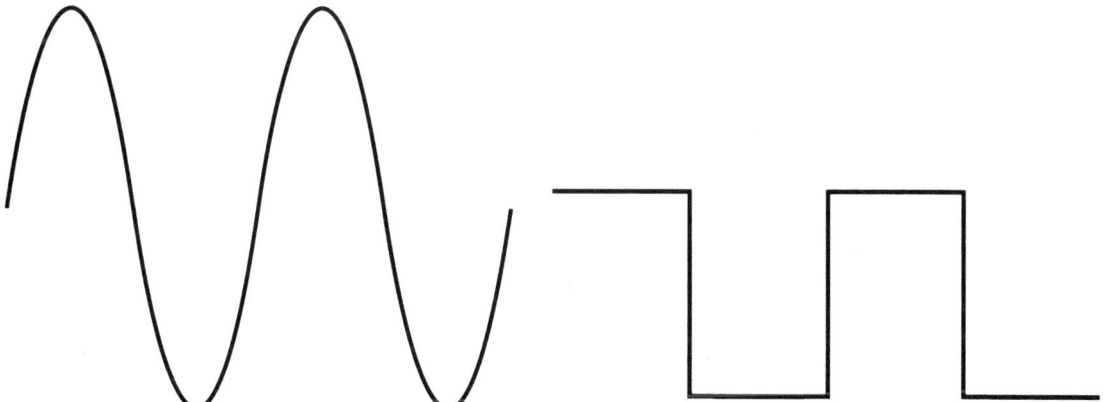

Figure 1-1. Sinusoidal waveform.   *Figure 1-2*. Square waveform.

4 / Digital Electronics

An *analog circuit* is an electronic circuit that generates an analog output waveform, or operates upon an analog input signal. An audio amplifier is an example of an analog circuit.

A *digital circuit* is an electronic circuit that generates a digital output waveform, or operates upon a digital input signal. A transistor switch is an example of a digital circuit.

The two states of a digital circuit may be identified in several ways. The states are often denoted as OFF and ON levels or states. Usually, the OFF state is the non-conducting level and the ON state is the conducting level. The OFF state is often referred to as the "0" level while the ON state is often referred to as the "1" level.

When the 1 level is more positive than the 0 level, the circuit is a positive logic device. When the 1 level is more negative than the 0 level, the circuit is a negative logic device.

INPUT	OUTPUT
0	1
1	0

Table 1-1. Truth table of an inverter.

SI	LI
OFF	OFF
ON	ON

Table 1-2. Truth table for *Example 1-1*.

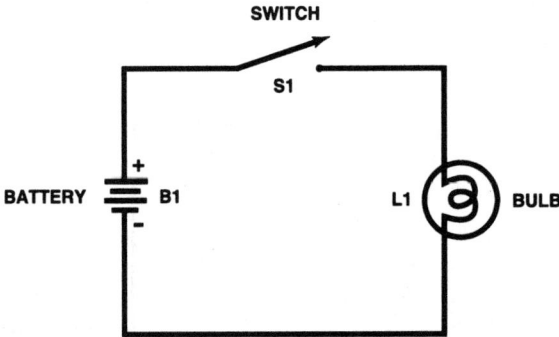

Figure 1-3. Circuit for *Example 1-1*.

The function of a digital circuit is often tabulated in the form of a function or *truth table*. All of the possible input states and the corresponding output states are tabulated. *Table 1-1* is the truth table of an inverter. The inverter will be discussed in the next chapter of this book.

Example 1-1:

Consider a light, a battery, and a switch connected as shown in *Figure 1-3*. When the switch is open or OFF, the light is off. The light is on when the switch is closed or ON. *Table 1-2* is the truth table for the circuit shown in *Figure 1-3*. This is a positive logic circuit because the ON state is more positive than the OFF state.

PROBLEMS

Problem 1-1.

Identify each of the following devices as either an analog or a digital device:

    (a) Switch
    (b) Relay
    (c) Faucet
    (d) Amplifier
    (e) Oscillator
    (f) Television
    (g) Dimmer light switch
    (h) Light bulb
    (i) Computer mainframe
    (j) Computer screen

Problem 1-2.

For the devices in question in *Problem 1-1*, identify those that are positive logic, negative logic, or neither positive nor negative logic. Explain why.

Problem 1-3.

Write a truth table for a relay. The input is the coil voltage, and the output is the relay contact. Assume that the relay contact is SPNO (single-pole, normally open).

Chapter Two
LOGIC GATES

A *logic gate* is an electronic circuit that performs a logic function. The simplest logic gate is the *inverter*, which is also known as the NOT gate. Other logic gates are the AND gate, the OR gate, and the XOR (exclusive-OR) gate.

INVERTER (NOT GATE)

The inverter changes a 1 level to a 0 level, and changes a 0 level to a 1 level. The operational symbol is $f = X'$, where f is the output level and X is the input level. *Figure 2-1* shows the symbols and the truth table for an inverter.

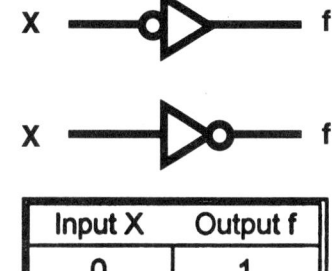

Figure 2-1. Truth table and symbol for an inverter.

The inverter can be drawn two ways. The circle indicates a 0 level. If the circle is at the input of the inverter, the gate is activated by a 0 level to produce a 1 level at the output. If the circle is at the output of the inverter, the gate is activated by a 1 level to produce a 0 level at the output. The output state is the opposite of the input state.

OR GATE

The interior light of an automobile is turned on either by opening one of the front doors OR by opening both front doors. This is the logic function of the OR gate. *Figure 2-2* is a schematic of an automobile interior light

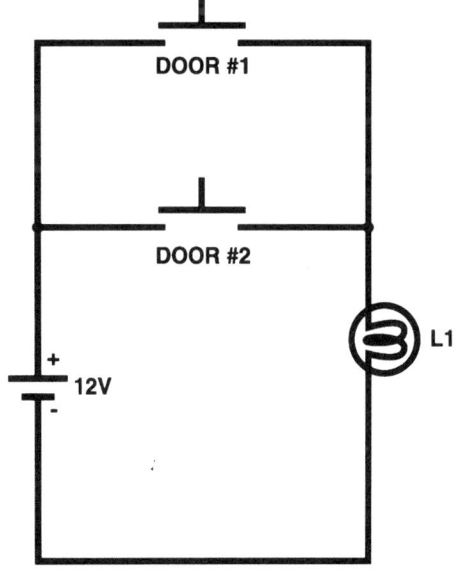

Figure 2-2. Schematic of OR type automobile light circuit.

circuit. The symbol of an OR gate and its truth table are shown in *Figure 2-3*. The operational symbol for an OR gate is f = X + Y, where X and Y are the inputs. The NOT OR gate is known as the NOR gate. It has the same function of an OR gate connected to an inverter. The NOR gate is represented by an OR gate with a circle on its output, as shown in *Figure 2-4*. The derivation of its truth table is left as an exercise for the reader. The output of an OR gate is LOW only when all of its inputs are LOW.

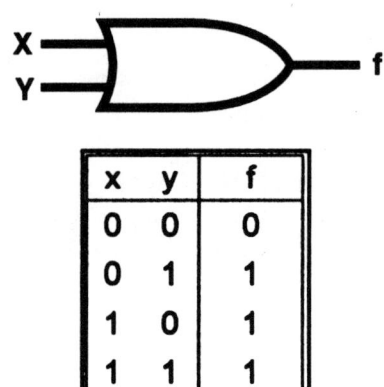

Figure 2-3. Truth table and symbol of an OR gate.

AND GATE

If a fuse is placed in series with the door switch of an automobile, both the fuse AND the switch must be closed for the interior light to turn on. If either the switch or the fuse is open, the light will not turn on. *Figure 2-5* is a schematic of the circuit. The symbol and truth table of an AND gate are shown in *Figure 2-6*. The AND function is written as f = X.Y, where X and Y are the inputs. The NOT AND gate, or NAND gate, is an AND gate with an inverter connected to its output. The NAND gate is denoted by a circle on its output, as shown in *Figure 2-7*. The derivation of its truth table is left as an exercise for the reader. The output of an AND gate is HIGH only when all of its inputs are HIGH.

XOR GATE

The XOR gate is a comparator. When the inputs are the same, the output is LOW. The output is HIGH when the inputs are different. The operational symbol is written as **f = X⊕Y**. The XOR function may be achieved by using AND gates because f = A'B + AB'. The XOR gate symbol and its truth table are shown in *Figure 2-8*.

PROBLEMS

Problem 2-1.
What is the truth table for a NOR gate?

Logic Gates / 9

Figure 2-4. Symbol of a NOR gate.

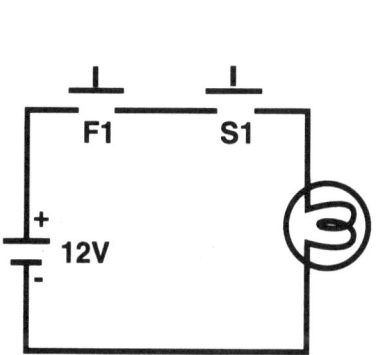

Figure 2-5. Schematic of an AND type automobile circuit.

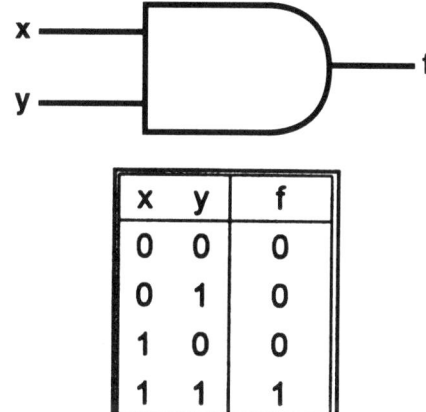

Figure 2-6. Truth table and symbol of an AND gate.

x	y	f
0	0	0
0	1	0
1	0	0
1	1	1

Figure 2-7. Symbol of an AND gate.

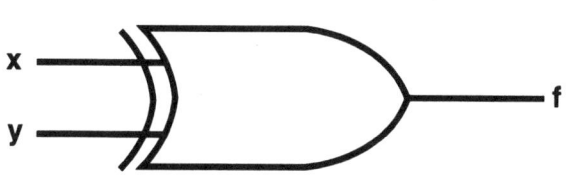

x	y	f
0	0	0
0	1	1
1	0	1
1	1	0

Figure 2-8. Truth table and symbol of an XOR gate.

10 / Digital Electronics

Problem 2-2.

What is the truth table for a NAND gate?

Problem 2-3.

What is the truth table of a XOR gate with an inverter connected to its output?

Problem 2-4.

Derive the truth table for the circuit shown in *Figure 2-9*.

Problem 2-5.

For the circuit shown in *Figure 2-10*, write a logic equation, derive its truth table, and draw its logic diagram.

Problem 2-6.

Write a logic equation to describe the circuit shown in *Figure 2-11*.

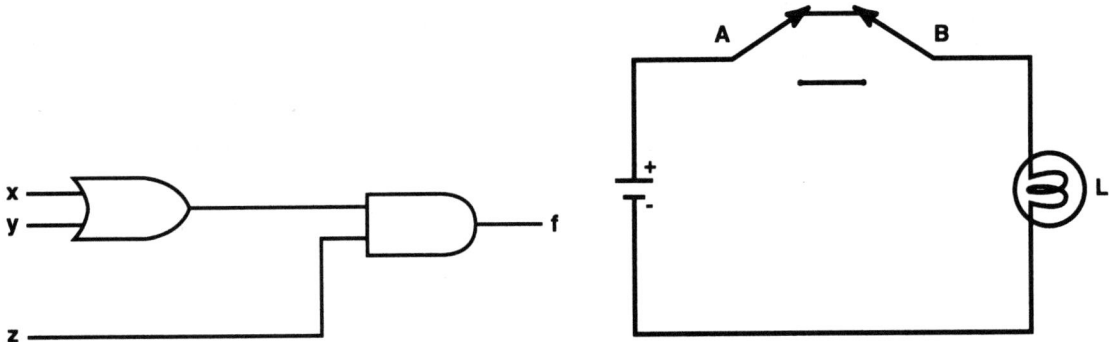

Figure 2-9. Circuit for *Problem 2-4*.   *Figure 2-10.* Circuit for *Problem 2-5*.

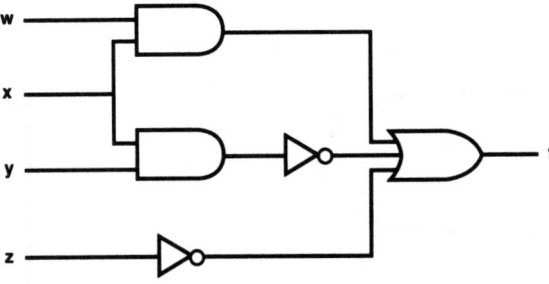

Figure 2-11. Circuit for *Problem 2-6*.

Chapter Three
LOGIC FAMILIES

It is necessary to define some logic gate performance characteristics before discussing logic families:

Fan-in is the number of inputs to a gate. A gate with three inputs is said to have a fan-in of three.

Current sourcing is when a current flows from the gate output into the load on that gate, as shown in *Figure 3-1*.

Current sinking is when a current flows from the load of a gate into the output of that gate, as shown in *Figure 3-2*.

Noise margin is the limit in the tolerance spread and the degree of loading of a gate. The transfer curve of the gate will never cross the shaded region, as shown in *Figure 3-3*. If the output loading of the gate is not exceeded:

    (a) The output voltage in the low state will never exceed V_{oL}.
    (b) The output voltage in the high state will never be less than V_{oH}.
    (c) Input voltages of V_{iH} or more guarantee a low level output of V_{oL} or less.
    (d) Input voltages of V_{iL} or less guarantee a high level output of V_{oH} or more.

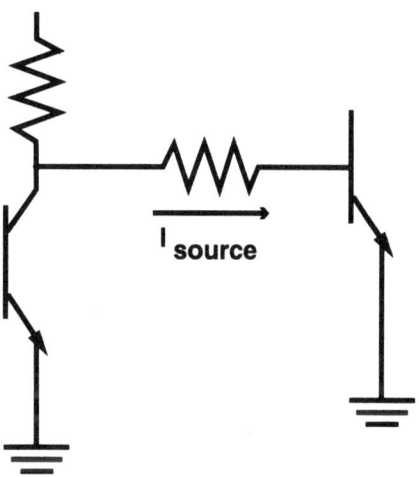

Figure 3-1. Current sourcing.

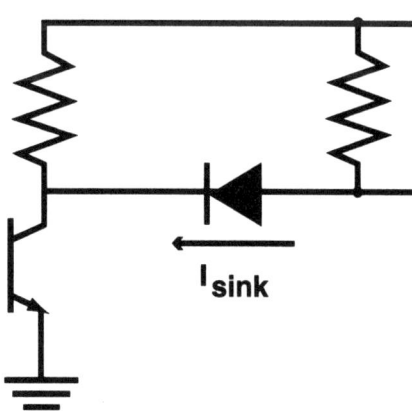

Figure 3-2. Current sinking.

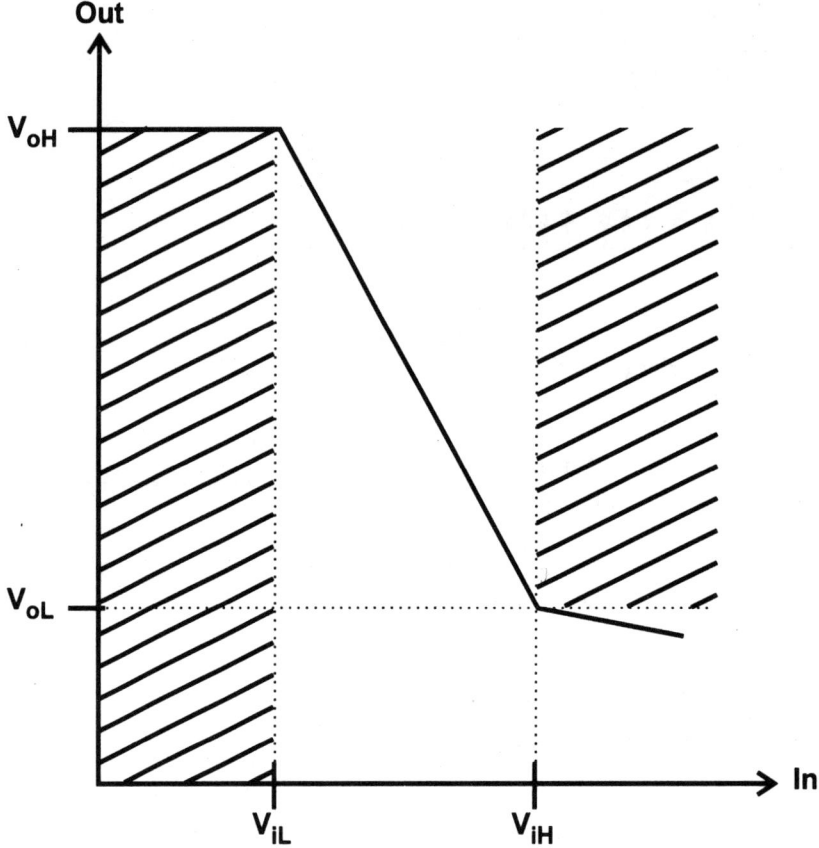

Figure 3-3. Transfer curve.

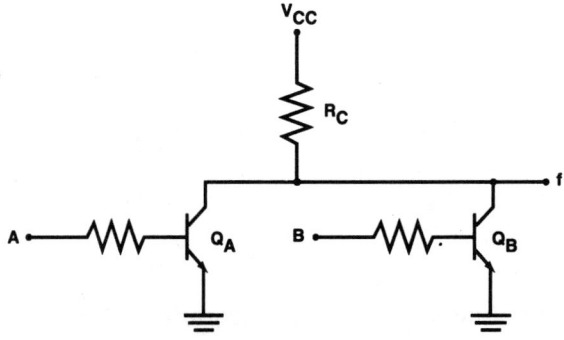

Figure 3-4. RTL gate.

Fan-out is the number of identical gates that a logic gate can drive before its output falls to the high-output lower limit, V_{oH}, for current sourcing logic; or rises to the low output upper limit, V_{oL}, for current sinking logic. A gate that can drive two identical gates is said to have a fan-out of two.

Propagation delay is the time it takes a signal to travel through a gate. Transistors cannot make instantaneous transitions between low- and high-output levels. The 0-to-1 and 1-to-0 propagation delays of a logic gate are different.

RESISTOR TRANSISTOR LOGIC (RTL)

RTL, or resistor transistor logic, was the earliest logic gate produced. *Figure 3-4* shows the schematic of an RTL gate. If the input to either transistor is high, current flows in RC and the output is low. The circuit is a NOR gate for positive logic and a NAND gate for negative logic. The characteristics of an RTL gate are:

Fan-in: Four or less.
Type: Current sourcing.
Noise immunity: Fair.
Fan-out: Poor.
Speed: Good for high power circuit, about 12 nsec. Fair for low power circuit, about 30 nsec.

DIODE TRANSISTOR LOGIC (DTL)

DTL, or diode transistor logic, uses diodes instead of resistors to control the base currents, as shown in *Figure 3-5*, which is a schematic of a DTL gate. When A and B are high, diodes D1 and D2 do not conduct; the transistor conducts because it

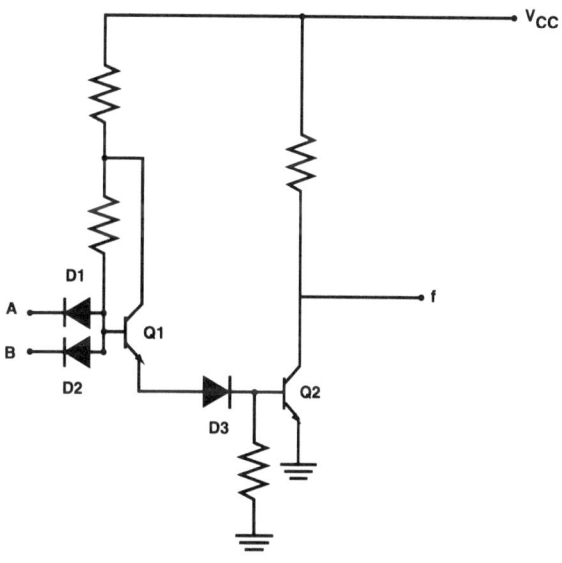

Figure 3-5. DTL gate.

receives base current through R1 and *f* is low. If either A or B goes low, the respective diode turns on and shunts to ground the base current of the transistor. The transistor turns off and *f* is high. The circuit is a NAND gate for positive logic and a NOR gate for negative logic.

The characteristics of a DTL gate are:

> *Fan-in*: Four or less; can be expanded with additional modules.
> *Type*: Current sinking.
> *Noise immunity*: Good.
> *Fan-out*: Good (typically eight).
> *Speed*: Good (25-80 nsec.).

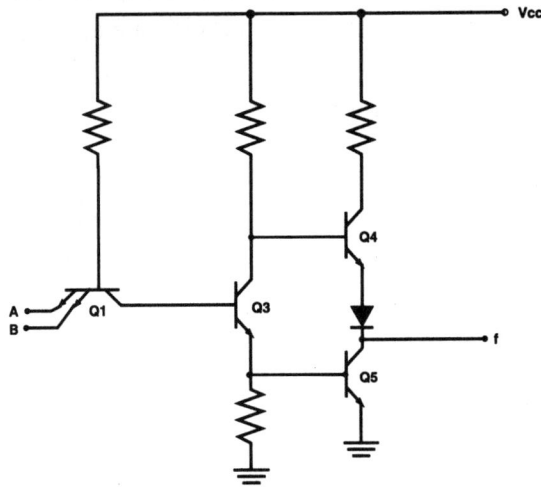

Figure 3-6. TTL gate.

TRANSISTOR-TRANSISTOR LOGIC (TTL)

In the DTL circuit, the input diodes are back-to-back with the offset diode, D3. This is essentially a transistor. The diodes are replaced by a transistor in TTL logic gates, as shown in *Figure 3-6*. Transistor Q4 is an active pull-up transistor, and it replaces RC. Therefore, it can source current to charge a line or load capacitance quickly. Transistor Q3 sinks current in the usual way. When A and B are high, the collector of Q1 and the base of Q3 are high. If either A or B is low, the collector of Q1 is low, the collector of Q2 is relatively high, and the base of Q3 is low. Q3 is off and Q4 can source current through a load to ground. The result is a positive logic NAND gate. The characteristics of TTL gates are:

> *Fan-in*: Eight.
> *Noise immunity*: Very good ($V_{oH} = 2.4V$, $V_{oL} = 0.4V$, $V_{iH} = 2V$, $V_{iL} = 0.8V$).
> $0NM = V_{iL} - V_{oL} = 0.8 - 0.4 = 0.4V$.
> $1NM = V_{oH} - V_{iH} = 2.4 - 2.0 = 0.4V$.
> *Fan-out*: Very good (typically 12).
> *Speed*: Highest for saturated circuitry, about 13 nsec.

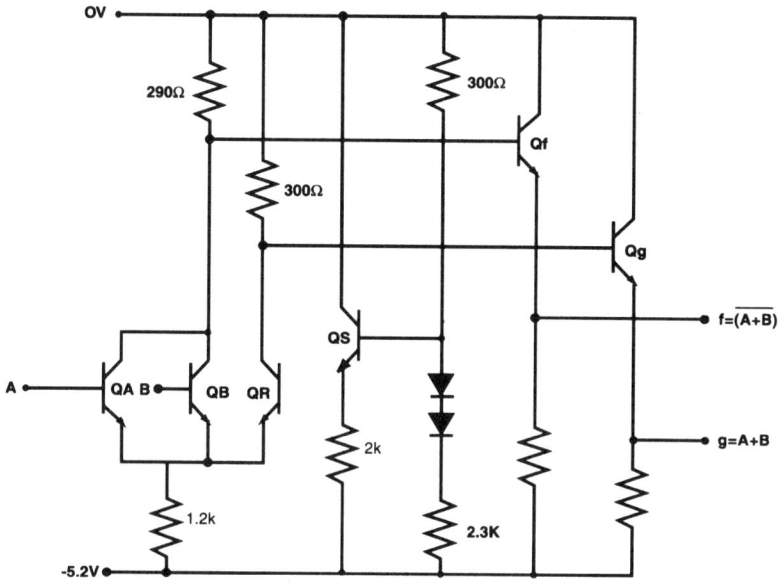

Figure 3-7. ECL gate.

EMITTER COUPLED LOGIC (ECL)

An ECL circuit cannot be saturated. It is either in its active region or in its cut-off region. The schematic of an ECL gate is shown in *Figure 3-7*.

The advantages of ECL are:
- (a) ECL operates at high speed because it cannot operate in its saturation region.
- (b) Noise margin is good because there is no internal noise generation, the high input impedance only requires low signal currents to be switched, and the complementary outputs allow balanced line interconnection.
- (c) Inverted and non-inverted outputs are available. This reduces chip count and also permits wired logic.
- (d) Fan-out is good at 25.

The disadvantages of ECL are its cost and high power consumption. Also, because of its speed, a transmission line-type interconnection is necessary.

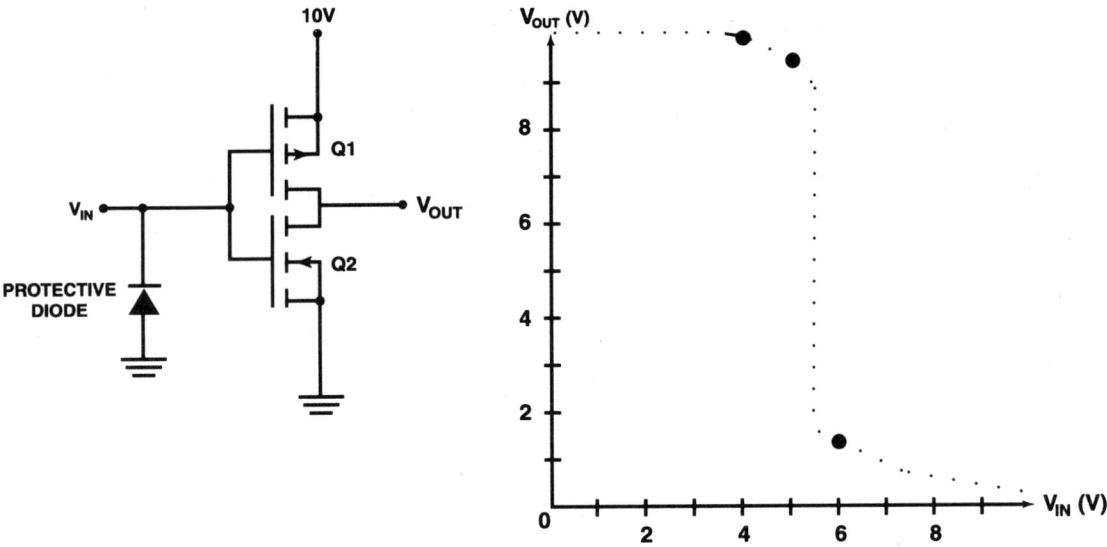

Figure 3-8. CMOS inverter and its transfer curve.

COMPLEMENTARY METAL-OXIDE SEMICONDUCTOR LOGIC (CMOS)

The CMOS family uses p-channel and n-channel enhancement mode MOSFET devices. A CMOS inverter and its transfer curve are shown in *Figure 3-8*. When the input signal exceeds 8 volts, Q1 is in its high resistance state and Q2 is in its low resistance state; the output is low. When the input signal is less than 2 volts, Q2 turns "off" and Q1 turns "on"; the output is high.

Power consumption is very low in CMOS circuits. The propagation delay of CMOS gates is greater than that of TTL gates. CMOS gates have high fan-out and high noise margins. CMOS gates use only active devices; therefore, a large amount of complex logic can be fabricated in a single integrated circuit. CMOS gates are slightly more expensive than TTL gates.

PROBLEMS

Problem 3-1.
Define fan-in, fan-out, current sourcing, current sinking, and noise margins.

Problem 3-2.

What is propagation delay? Why is there a propagation delay?

Problem 3-3.

What is the earliest logic family?

Problem 3-4.

Calculate the 0NM and the 1NM for a logic family with the following specifications:

$$V_{oH} = 10V, V_{oL} = 1V, V_{iH} = 8V, V_{iL} = 2V.$$

Problem 3-5.

Which logic family cannot enter the saturation state?

Problem 3-6.

Which logic family is ideal for battery-powered circuits? Why?

Problem 3-7.

Compare the characteristics of the logic families.

Chapter Four
LOGIC FUNCTION IMPLEMENTATION

The first significant contribution to switching theory was made in 1938 by C.E. Shannon. He developed the algebra of switching circuits and demonstrated its relationship to the calculus of propositions and Boolean algebra. Shannon made further contributions to switching theory during the 1940s and the 1950s.

Boolean algebra is a two-valued algebra that may be used in switching theory by assigning 0 and 1 to the two voltage levels upon which most switching networks are designed.

The Boolean connective "+" is read as OR, and is defined in *Table 4-1*. The Boolean connective "." is read as AND, and is defined in *Table 4-2*. If the dot is absent, it is still read as AND.

A	B	f = A + B
0	0	0
0	1	1
1	0	1
1	1	1

Table 4-1. Truth table for an OR gate.

Some useful theorems in the field of switching theory are:

(a) $A.0 = 0$
(b) $A.1 = A$
(c) $A.A = A$
(d) $A.A' = 0$
(e) $A + 0 = A$
(f) $A + 1 = 1$
(g) $A + A = A$
(h) $A + A' = 1$
(i) $A'' = A$
(j) $A.B = B.A$
(k) $A + B = B + A$
(l) $A.B + A.C = A(B + C)$
(m) $(A + B)(A + C) = A + (B.C)$
(n) $A.B + A.B' = A$

A	B	f = AB
0	0	0
0	1	0
1	0	0
1	1	1

Table 4-2. Truth table for an AND gate.

Theorems (a) to (d) can be verified by letting A = 0 and A = 1, and substituting into *Table 4-1*. Theorems (e) to (h) can be verified by letting A = 0 and A = 1, and substituting into *Table 4-2*. Theorem (i) states that a double inverse cancels out. Theorems (j) and (k) state that the order of the variables is unimportant; they are therefore commutative. Theorems (l) and (m) are factoring theorems. Theorem (n) is useful for minimizing logic circuits.

SUM OF PRODUCTS

The sum of products is written as: f = A.B + C.D = AB + CD. The circuit using AND and OR gates is shown in *Figure 4-1*. The AND gate provides the *product function*, and the OR gate provides the *sum function*.

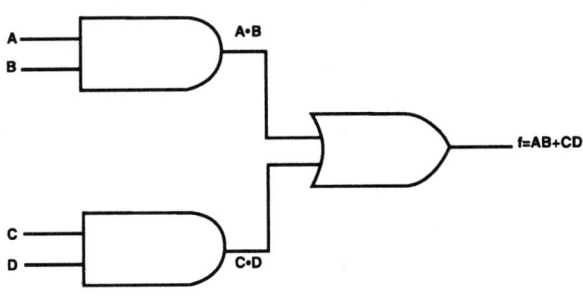

Figure 4-1. Circuit for the sum of products.

PRODUCT OF SUMS

The product of sums is written as: f = (A + B).(C + D) = (A + B)(C + D). The circuit using OR and AND gates is shown in *Figure 4-2*. The OR gate provides the sum function, and the AND gate provides the product function.

WIRED LOGIC

Wired logic is positive logic, and is used to reduce the number of gate leads required for a logic implementation. Wired logic yields the inverted function (f') output. Therefore, one should determine the f' required because f" = f, according to theorem (i).

Example 4-1

f = X'Y'.W'V' = $\overline{XY}.\overline{WV}$ = $\overline{XY + WV}$, the schematic of which is shown in *Figure 4-3*. The reader should not that X' = $\overline{X}$.

KARNAUGH MAP

The Karnaugh map is a Venn diagram, and it provides an alternative to the truth table for defining a logic function. The output state corresponding to each combination of variables is entered into a particular cell of the Karnaugh map array. In a Karnaugh map, two, four, eight, sixteen, etc., adjacent cells with the same output function may be grouped together to form a single area. The simplest form of a functional expression may thus be obtained.

Figures 4-4 and *4-5* show the truth tables and the Karnaugh map for a two-variable and three-variable function, respectively. The next few examples illustrate how to use a Karnaugh map to implement a logic function.

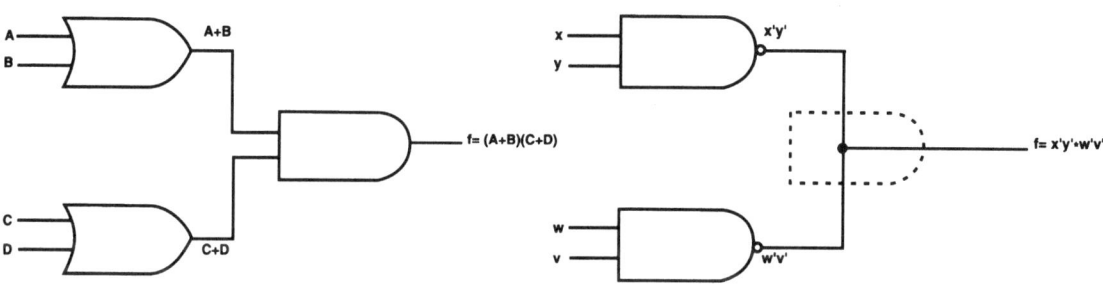

Figure 4-2. Circuit for the product of sums.

Figure 4-3. Schematic of the function of *Example 4-1*.

A	B	f
0	0	f_0
0	1	f_1
1	0	f_2
1	1	f_3

B \ A	0	1
0	f_0	f_2
1	f_1	f_3

Figure 4-4. Truth table and Karnaugh map for a two-variable function.

A	B	C	f
0	0	0	f_0
0	0	1	f_1
0	1	0	f_2
0	1	1	f_3
1	0	0	f_4
1	0	1	f_5
1	1	0	f_6
1	1	1	f_7

A	0	0	1	1
B	0	1	1	0
C = 0	f_0	f_2	f_6	f_4
C = 1	f_1	f_3	f_7	f_5

Figure 4-5. Truth table and Karnaugh map for a three-variable function.

Example 4-2

It is required to implement the truth table given in *Table 4-3*. *Figure 4-6* is the Karnaugh map for *Table 4-3*. All of the "1"s are grouped together.

f = ABC + ABC' + A'BC + A'BC' = AB(C + C') + A'B(C + C') = AB + A'B = B(A + A') = B

Therefore, f = B using Theorems (h) and (l).

Example 4-3

It is sometimes simpler to implement the logic function by mapping for the "0"s. *Figure 4-7* shows the Karnaugh map for *Table 4-4*. If the reader maps for the zeros of f, the reader obtains f' = AC' + A'BCD. If the reader maps for the ones of f', the reader once again obtains f' = AC' + A'BCD.

A	B	C	f
0	0	0	0
0	0	1	0
0	1	0	1
0	1	1	1
1	0	0	0
1	0	1	0
1	1	0	1
1	1	1	1

Table 4-3. Truth table for *Example 4-2*.

Table 4-6. NOR and NAND equivalents for elementary logic gates.

Logic Function Implementation / 23

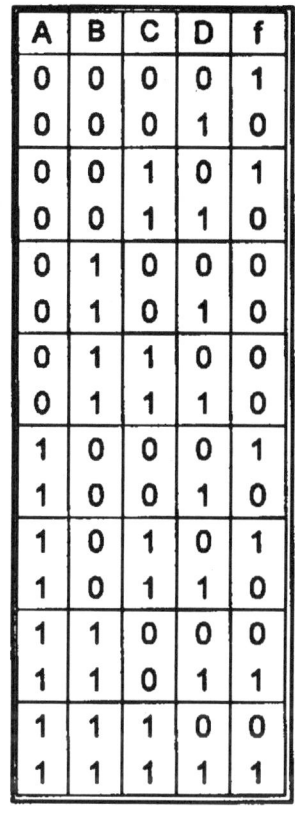

A	B	C	D	f	f̄
0	0	0	0	1	0
0	0	0	1	1	0
0	0	1	0	1	0
0	0	1	1	1	0
0	1	0	0	1	0
0	1	0	1	1	0
0	1	1	0	1	0
0	1	1	1	0	1
1	0	0	0	0	1
1	0	0	1	0	1
1	0	1	0	1	0
1	0	1	1	1	0
1	1	0	0	0	1
1	1	0	1	0	1
1	1	1	0	1	0
1	1	1	1	1	0

Table 4-4. Truth table for *Example 4-3*.

A	B	C	D	f
0	0	0	0	1
0	0	0	1	0
0	0	1	0	1
0	0	1	1	0
0	1	0	0	0
0	1	0	1	0
0	1	1	0	0
0	1	1	1	0
1	0	0	0	1
1	0	0	1	0
1	0	1	0	1
1	0	1	1	0
1	1	0	0	0
1	1	0	1	1
1	1	1	0	0
1	1	1	1	1

Table 4-5. Truth table for *Example 4-4*.

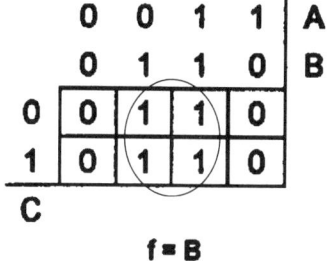

Figure 4-6. Karnaugh map for *Example 4-2*.

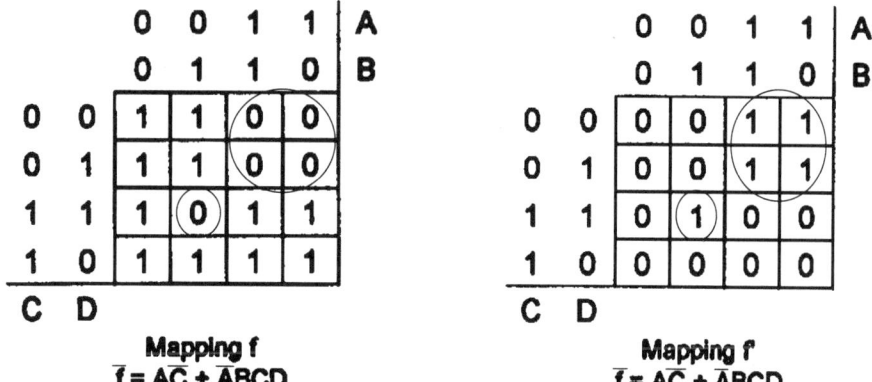

Figure 4-7. Karnaugh maps for *Example 4-3*.

The four-variable Karnaugh map may be considered as a flattened toroid. The two vertical boundaries and the two horizontal boundaries are adjacent.

Example 4-4

It is required to implement the logic function of *Table 4-5*. The four corners may be considered as one area, as illustrated in *Figure 4-8*. Therefore, f = B'D' + ABD.

ANALYZING MULTILEVEL CIRCUITS

DeMorgan's theorem permits the implementation of Boolean algebra expressions using inverted logic NOR and NAND gates. The dual forms of the theorem are:

(a) $\overline{XY} = \overline{X} + \overline{Y}$, and is diagrammatically shown in *Figure 4-9*.
(b) $\overline{X+Y} = \overline{X}\,\overline{Y}$, and is diagrammatically shown in *Figure 4-10*.

Each theorem may be proved by showing the equivalence of the two sides of each equation by using either truth tables or Karnaugh maps.

When all of the inputs of a NAND or NOR gate are connected together, the gate acts as an inverter. *Table 4-6* summarizes the NOR and NAND equivalents of the elementary logic gates.

Example 4-5

Draw the NAND gate circuit for f = A'B' + A'C' + A'D' + AC. The NAND gate circuit is shown in *Figure 4-11* and was obtained by using *Table 4-6*.

Example 4-6

Repeat *Example 4-5* using NOR gates. The NOR gate circuit is shown in *Figure 4-12* and was obtained by using *Table 4-6*.

It is sometimes necessary to simplify multilevel circuits. The procedure is:
  (a) Starting with the output, mark off each level of the circuit.
  (b) Replace each odd-numbered level NAND and/or NOR gate with its negated input equivalent, according to deMorgan's theorem.
  (c) Cancel "0"s or circles according to Theorem (i). The resulting circuit is the simplified circuit required.

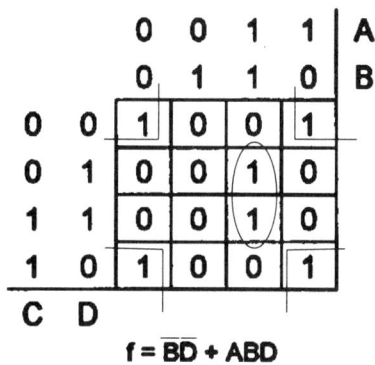

$f = \overline{BD} + ABD$

Figure 4-8. Karnaugh map for *Example 4-4*.

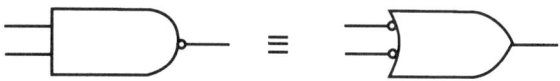

Figure 4-9. Diagram for the De Morgan theorem I.

Figure 4-10. Diagram for the De Morgan theorem II.

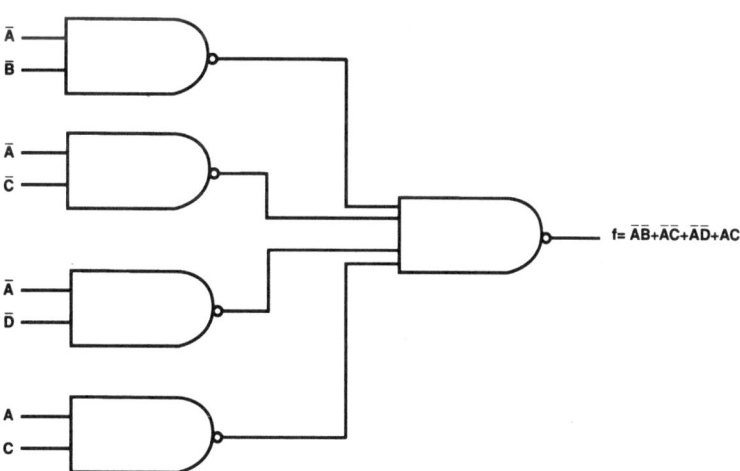

Figure 4-11. NAND gate circuit for *Example 4-5*.

Example 4-7

Simplify the multilevel circuit shown in *Figure 4-13*. *Figure 4-14* shows the final circuit and the steps required to get the simplified circuit.

PROBLEMS

Problem 4-1.

Verify theorems (a), (b), (c), and (d).

Problem 4-2.

Verify theorems (e), (f), (g), and (h).

Problem 4-3.

Prove theorem (n).

Problem 4-4.

Draw the Karnaugh map for a 2-input AND gate.

Problem 4-5.

Draw the Karnaugh map for a 3-input OR gate.

Problem 4-6.

Draw the Karnaugh map for a 4-input NAND gate.

Problem 4-7.

Draw the Karnaugh map for a four-variable function.

Problem 4-8.

Draw the Karnaugh map for a five-variable function.

Problem 4-9.

Implement the logic function for the truth table shown in *Table 4-7* using AND and OR gates.

Problem 4-10.

Repeat *Problem 4-9* using NAND gates.

Logic Function Implementation / 27

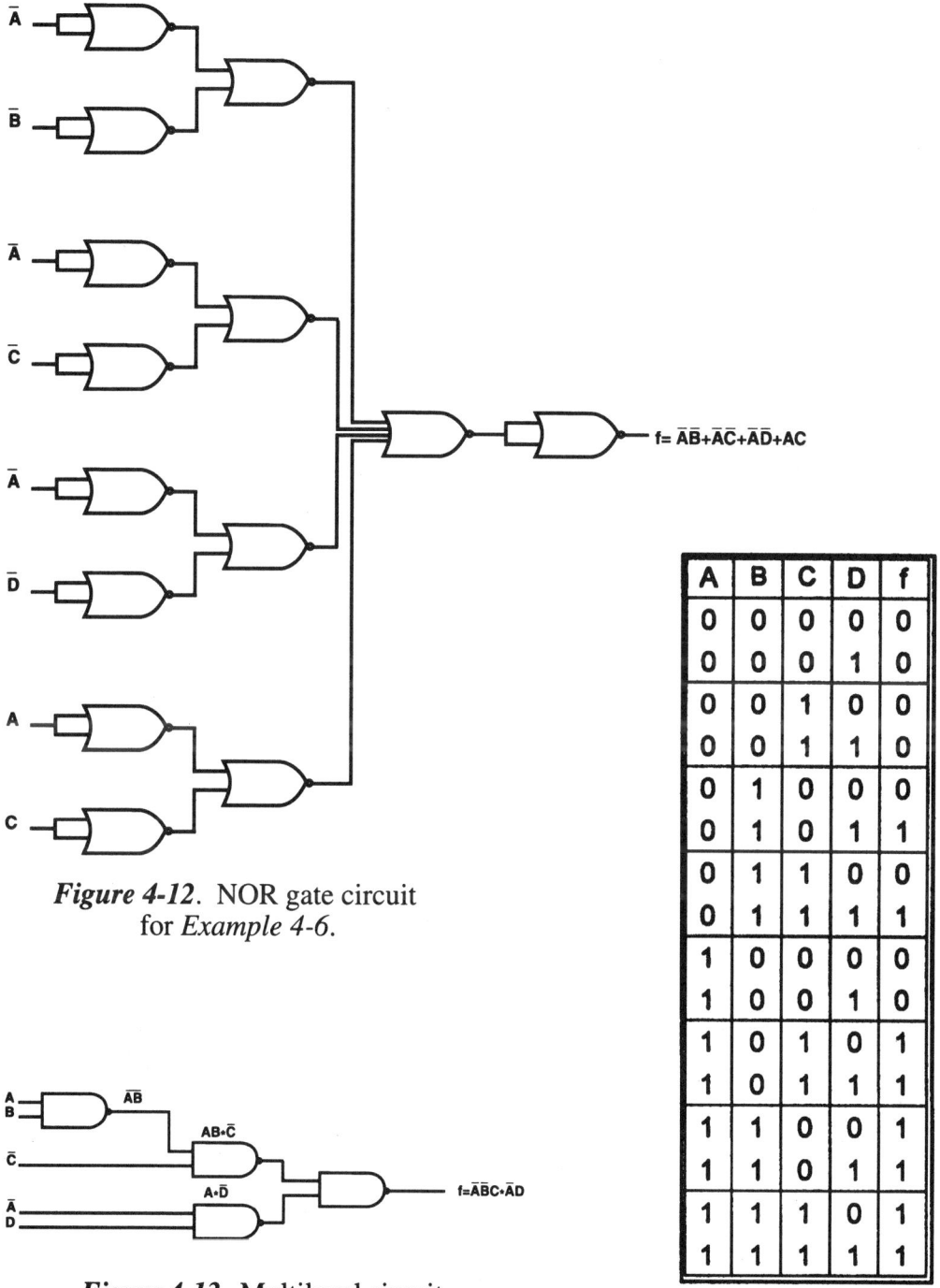

Figure 4-12. NOR gate circuit for *Example 4-6*.

Figure 4-13. Multilevel circuit of *Example 4-7*.

Table 4-7. Truth table for *Problem 4-9*.

Problem 4-11.

Repeat *Problem 4-9* using NOR gates.

Problem 4-12.

Prove deMorgan's theorem.

Problem 4-13.

Simplify the multilevel logic circuit shown in *Figure 4-15*.

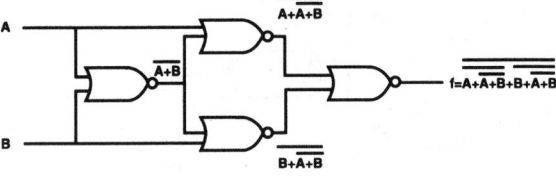

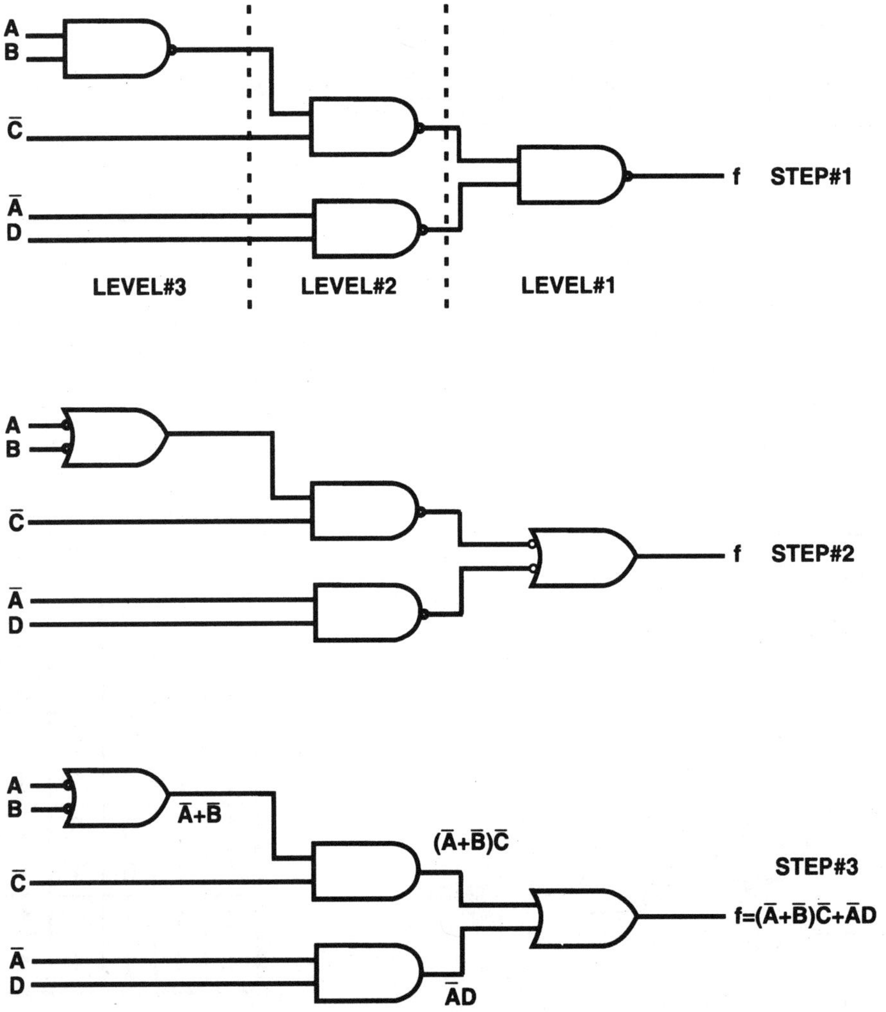

Figure 4-15. Multilevel logic circuit for Problem 4-13.

Figure 4-14. Simplified circuit for *Example 4-7*.

Chapter Five
FLIP-FLOPS

In logic gate circuits, the output is determined only by the inputs existing at that time. *Figure 5-1* shows the schematic of a NOR gate latch where the output is 1 when the start input is a momentary 1. The truth table of the NOR gate latch is also shown in *Figure 5-1*. The output is 0 when a momentary 1 is applied to the stop input. The schematic and truth table for the NAND gate latch are shown in *Figure 5-2*.

The output of a latch is dependent upon the previous inputs. The latch can be considered to have a *memory*.

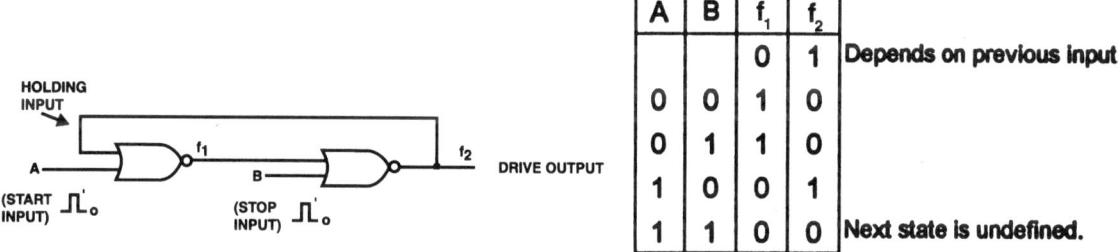

Figure 5-1. NOR gate latch schematic and truth table.

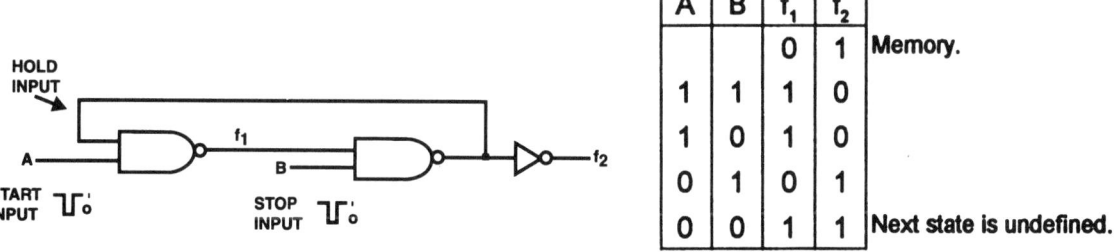

Figure 5-2. NAND gate latch schematic and truth table.

FLIP-FLOP

The latch circuit may be redrawn differently, as shown in *Figure 5-3*. The circuit is known as a Reset-Set flip-flop, or an RS latch. The circuit is used to store a single binary digit because it will remain in one of its two states until it is intentionally disturbed.

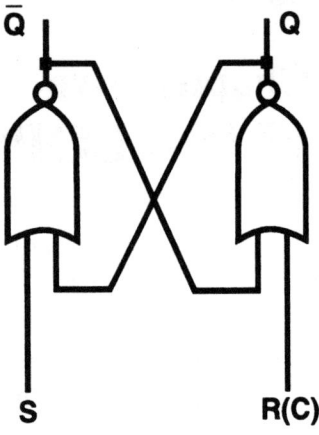

Figure 5-3. RS flip-flop.

The outputs are labeled Q and $\bar{Q}$, or the 1 and 0 outputs, respectively. When $Q=0$ and $\bar{Q}=1$, the flip-flop is reset; that is, it stores a 0. When $Q=1$ and $\bar{Q}=0$, the flip-flop is set; that is, it stores a 1.

A momentary 1 on the *S* or *Set* input will set the flip-flop, while a momentary 1 on the *R* or *Reset* input will reset or clear the flip-flop. *Figure 5-4* shows the schematic and truth table for the NOR and NAND gate RS latch.

CLOCKED AND GATED FLIP-FLOP

The RS latch is susceptible to spurious triggering and therefore cannot be used in complex logic systems. Small timing differences between the signals to a logic gate can result in transient outputs that can initiate an unwanted change of state in the latch circuit. *Figure 5-5* shows how timing differences can generate a transient output.

To eliminate the problem, the latch inputs are applied through AND gates (or equivalent) which in turn are conditioned by a timing clock pulse. The flip-flop is therefore disconnected from its inputs except during a clock pulse. *Figure 5-6* shows the schematic and truth table for a clocked and gated flip-flop. The timing diagram of a clocked and gated flip-flop is shown in *Figure 5-7*. The reader should note that Q goes high when both the clock pulse and S are high. $\bar{Q}$ goes high when both the clock pulse and R are high.

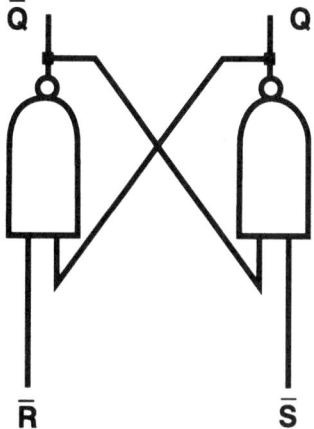

$\bar{S}$	$\bar{R}$	Q	$\bar{Q}$	
0	0	1	1	Outputs undefined; Inputs removed at same time.
1	0	1	0	Active input = 0.
0	1	0	1	
1	1	1	0	Depends on last input (latched condition).
1	1	0	1	

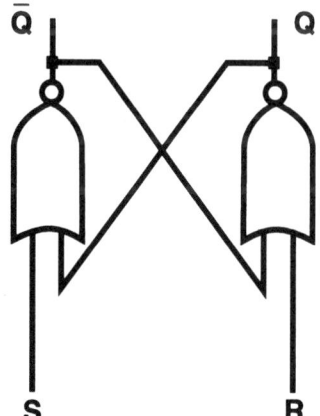

S	R	Q	$\bar{Q}$	
0	0	0	1	Depends on the last input (latched condition).
0	0	1	0	
0	1	0	1	Active input = 1.
1	0	1	0	
1	1	0	0	Outputs undefined if inputs removed at the same time.

Figure 5-4. NOR and NAND gate RS latch.

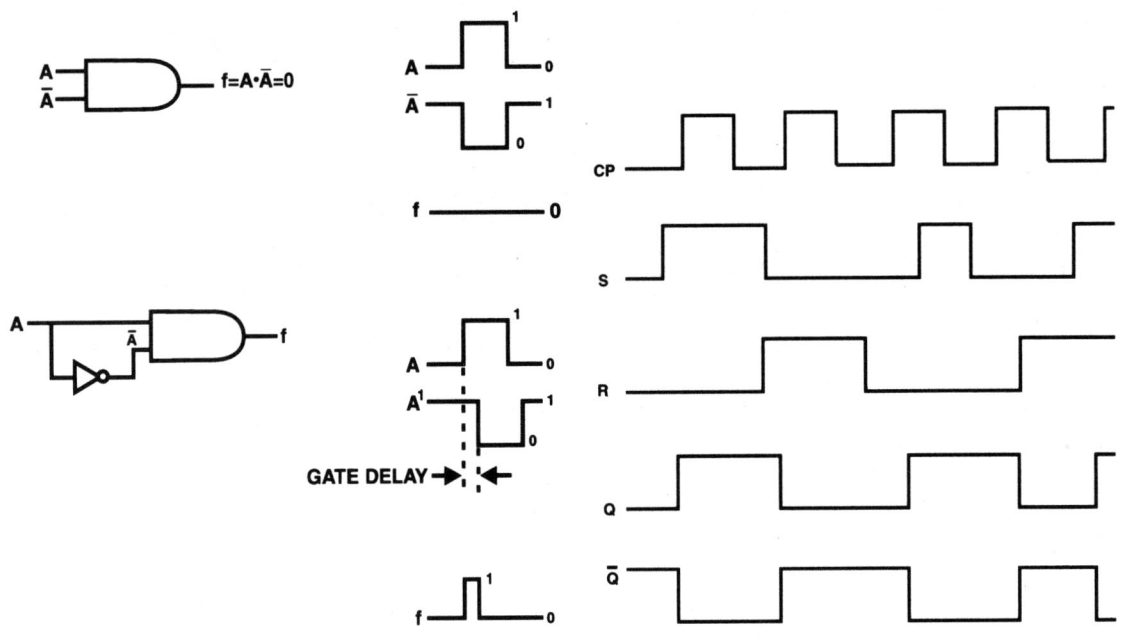

Figure 5-5. Generation of a transient output.

Figure 5-7. Timing diagram for clocked and gated flip-flop.

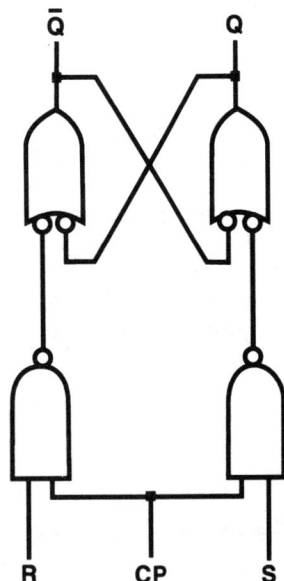

S	R	Q	$\bar{Q}$
0	0	No Change	No Change
0	1	0	1
1	0	1	0
1	1	1	1

While clock is high. Undefined when clock goes low

Figure 5-6. Clocked and gated flip-flop schematic and truth table.

MASTER-SLAVE FLIP-FLOP

It is possible for both the set and clear inputs to change during the same clock pulse. The flip-flop would respond to both input changes and the output would be uncertain. This problem can be eliminated only if the inputs can change on the leading edge (transition from LOW to HIGH) of the clock pulse, or on the trailing edge (transition from HIGH to LOW) of the clock pulse. This can be achieved by a master-slave configuration, as shown in *Figure 5-8*.

The timing diagram for the master-slave flip-flop is also shown in *Figure 5-8*. On the leading edge of the clock pulse, the slave latch is disconnected from the master latch by the inverted clock pulse and the master latch accepts the data input indicated by the set and clear inputs.

On the trailing edge of the clock pulse, the master latch is disconnected from the inputs, the slave latch is reconnected to the master latch by the rising inverted clock pulse, and the state of the master latch is transferred to the slave latch.

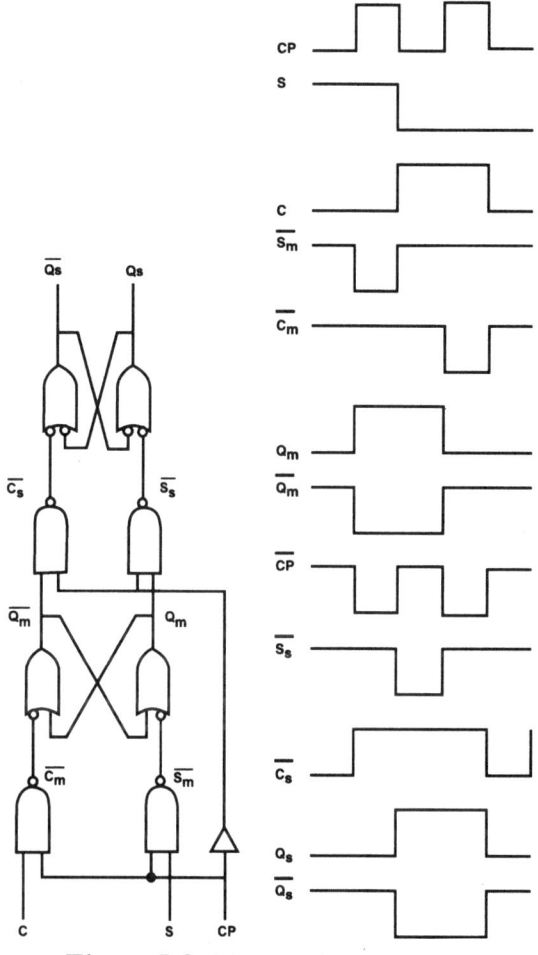

Figure 5-8. Master-slave flip-flop circuit and timing diagram.

DIRECT CLEAR AND DIRECT SET INPUTS

The direct clear and direct set inputs are also known as *asynchronous inputs*. The direct clear can be used for returning the flip-flop to a 0 state independently of the clock pulse; that is, asynchronously. The direct set permits asynchronous resetting of the flip-flop to a 1 state; that is, independently of the clock pulse. In the master-slave arrangement, the direct set and clear inputs act directly on the slave latch as well as the master latch.

Additional set and clear inputs are sometimes available so that the switching of the flip-flop can be conditioned without additional external gating. A typical clocked and gated RS flip-flop is shown in *Figure 5-9*. The circles on the S_d and C_d inputs indicate that asynchronous setting or clearing of the flip-flop is achieved by a 0 level. For normal synchronous operation, these inputs should be held at a 1 level.

D FLIP-FLOP

The D flip-flop has an inverter between the set and clear inputs to ensure that these inputs are different. The D flip-flop can be used as a one clock interval time delay or as a single line data entry. The schematic, truth table, and timing diagram of a D flip-flop are shown in *Figure 5-10*.

JK FLIP-FLOP

The JK flip-flop has its outputs and inputs connected as shown in *Figure 5-11*. This removes the restriction that both inputs cannot be 1. When $J = K = 1$, the JK flip-flop reverses state when it is clocked. The truth table for the JK flip-flop is also shown in *Figure 5-11*.

PROBLEMS

Problem 5-1.
Draw the equivalent logic circuit of a clocked and gated flip-flop using (a) only NOR gates, and (b) only NAND gates. What happens if both logic inputs are held at the enabling level while the circuit is clocked? Write a truth table for each circuit.

Problem 5-2.
Draw the equivalent NAND gate logic circuit of a master-slave flip-flop. Draw its timing diagram. Repeat using NOR gates.

Problem 5-3.
What is meant by "direct set" and "direct clear" in a flip-flop? How is this achieved in a master-slave flip-flop?

Problem 5-4.
Write a truth table for a clocked RS flip-flop and explain what it means.

Flip Flops / 35

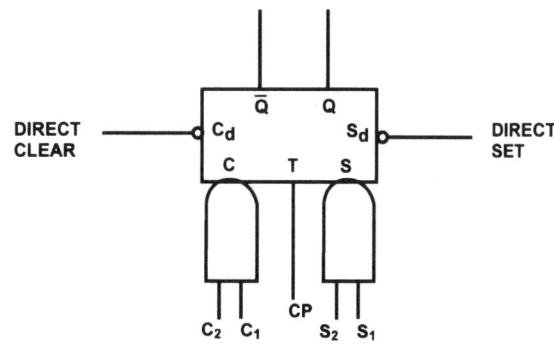

Figure 5-9. Clocked and gated RS flip-flop.

Time t_n (Before clock)		Time t_{n+1} (After clock)
C	S	Q_{n+1}
0	0	Q_n (No change)
0	1	1 (After clock)
1	0	0 (After clock)
1	1	X

X = Indeterminate and not allowed.

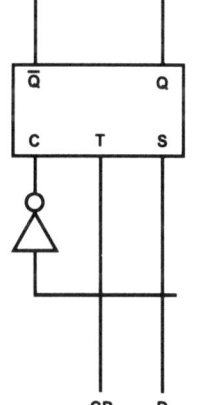

t_n	t_{n+1}
D	Q_{n+1}
0	0
1	1

Figure 5-10. Schematic, truth table, and timing diagram for a D flip-flop.

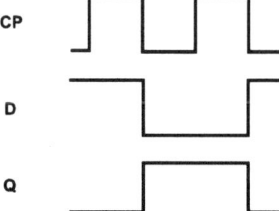

Problem 5-5.

For the circuit of *Figure 5-12*, given the input A, draw the waveforms at B, C, and D. All flip-flops are initially reset.

Problem 5-6.

Draw the equivalent logic circuit for a D flip-flop. Redraw the circuit of *Figure 5-12* using only D flip-flops.

Problem 5-7.

Draw the equivalent logic circuit and write a truth table for a JK flip-flop. How does the JK flip-flop differ from the clocked RS flip-flop?

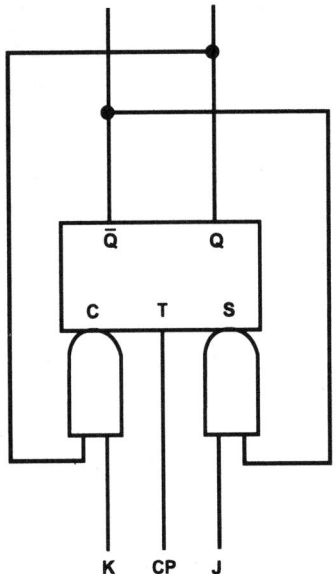

	t_n		t_{n+1}	
J	K	Q_{n+1}		
0	0	Q_n	No change.
0	1	0	
1	0	1	
1	1	$\overline{Q_n}$	Toggle.

Figure 5-11. JK flip-flop schematic and truth table.

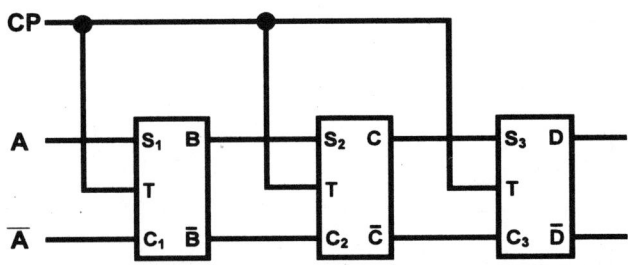

Figure 5-12. Circuit for *Problem 5-5*.

Chapter Six
CONTROL CIRCUITS

A control circuit turns a device (such as a motor or lamp) on or off in response to a specific set of input conditions. The input conditions must be binary in nature. Input conditions are usually described by the opening or closing of a switch, or the change of state of a transducer output as the physical parameter (such as temperature or pressure) varies around a prescribed threshold point. Some input devices that are frequently used are switches, thermostats, photocells, etc. The output level of a logic circuit cannot drive a device that requires significant power. A power transistor switch is often used to drive a load. A Darlington pair amplifier can also be used as an interface between a control circuit and a heavy load.

A combinational logic circuit may be designed using a truth table or a Karnaugh map. Very simple combinational logic circuits may be designed by intuition.

Designing a combinational circuit is quite easy. A truth table is drawn in which a column is used for each input and each output. All the possible input states are listed, and the output function is denoted by a 1 if it is to occur, or by a 0 if it is not to occur. The truth table is redrawn as a Karnaugh map where adjacent ones are mapped together to yield a minimized output function.

Example 6-1

A three-stage air conditioner functions with two compressors. The smaller compressor, M1, operates when the temperature is between 75° and 85°F. The larger compressor functions when the temperature is between 85° and 95°F. Both compressors operate at temperatures above 95°F. A NOR logic control circuit is required. The design process and the final control circuit are shown in *Figure 6-1*.

Sometimes a control circuit is required to operate with only momentary input signals. This type of control circuit requires a latch circuit to provide a continuous drive output. There are several input signals for this type of circuit.

38 / *Digital Electronics*

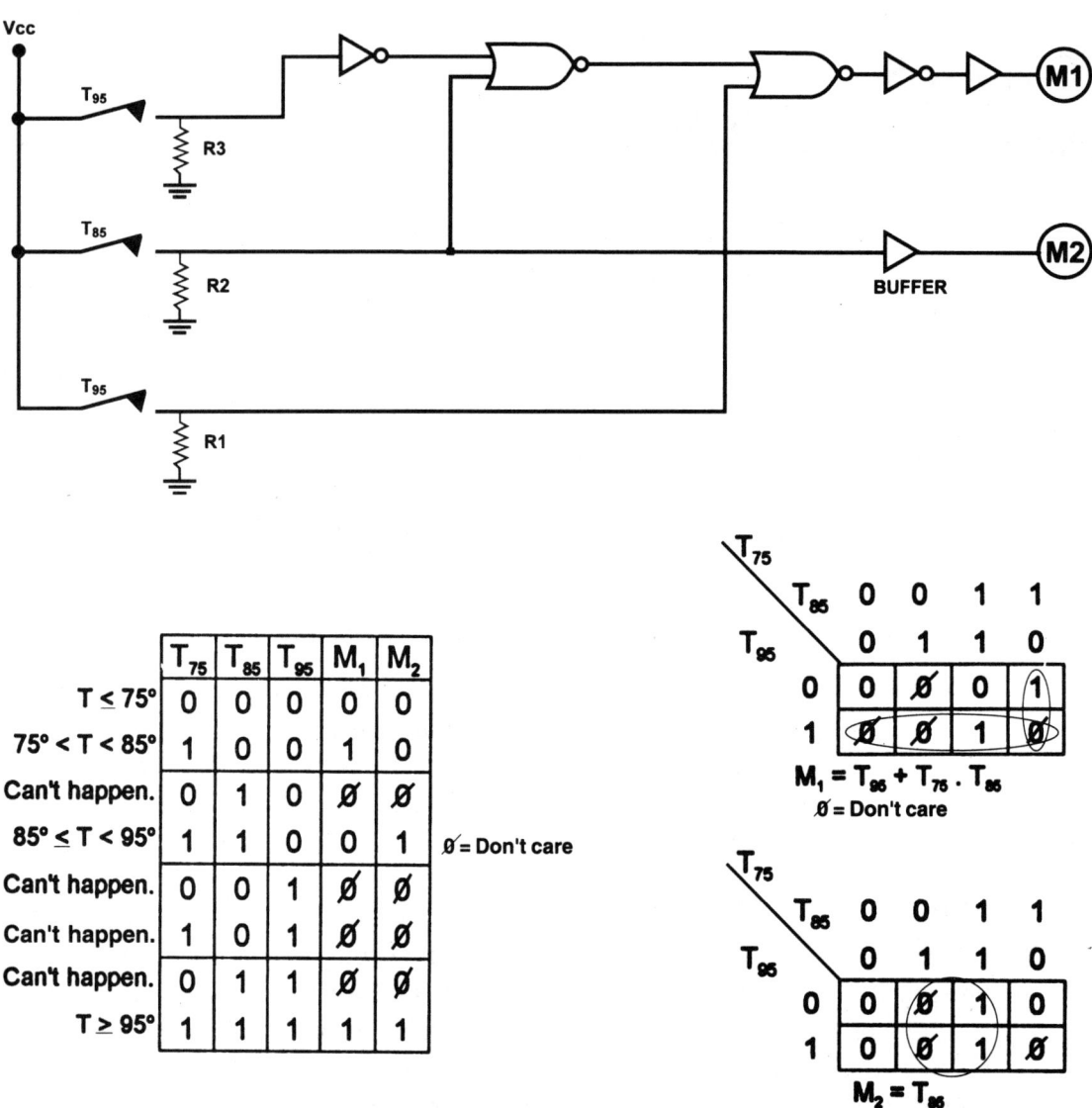

Figure 6-1. Design and circuit for *Example 6-1*.

Control Circuits / 39

The START and STOP signals are separate so that if both are pushed simultaneously, the latch turns off. The START CONDITION or S/C must be in a specific state before the operation of the START input can be effective. The OVERRIDE or O/R input overrides the momentary start pulse; or if the control circuit is already energized, it turns off the drive output. The INTERRUPT or INT input temporarily interrupts the output without clearing the latch.

The START and START CONDITION inputs are "ANDed" together, while the STOP and OVERRIDE inputs are "ORed" together. The INTERRUPT input is gated with the drive output.

The NOR and NAND sequential control circuits are shown in *Figure 6-2*.

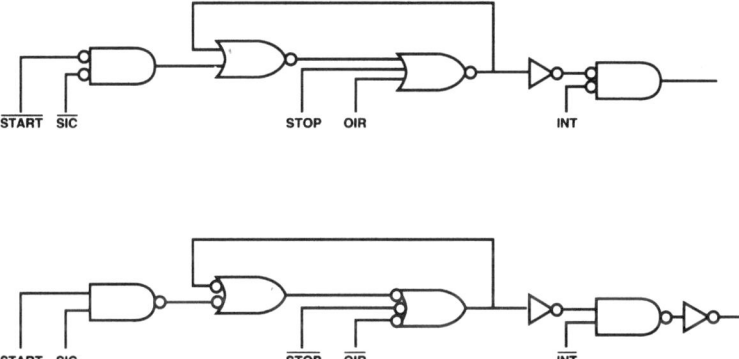

Figure 6-2. NOR and NAND sequential control circuits.

Example 6-2

A motor relay is energized when a momentary start button, A, is pressed and if the limit switches, L1 and L2, are closed. The relay is to drop out if a momentary stop button, B, is pressed or if limit switch L2 opens. When the relay is energized, limit switch L1 has no effect on it. Design the required logic using NAND gates.

Only one output and one latch are required:

    START: A
    S/C: L1
    STOP: B
    O/R: $\overline{L2}$
    INT: -

It should be noted that L2 is not a START CONDITION. It is an OVERRIDE condition because it can stop operation of the control circuit once it is energized. The NAND sequential control circuit is shown in *Figure 6-3*.

PROBLEMS

Problem 6-1.

A heating system consists of an electric furnace with two heating coils, C1 and C2, and a heat pump. When the temperature is above 0°F, the heat pump can supply all of the heating requirements. When the temperature is between 0° and -15°F, the heat pump and the smaller heating coil, C1, supply the required heat. When the temperature is less than -15°F, the larger heating coil supplies the heat required. Design a NOR logic control circuit.

Problem 6-2.

Design a sequential control circuit using an RS flip-flop.

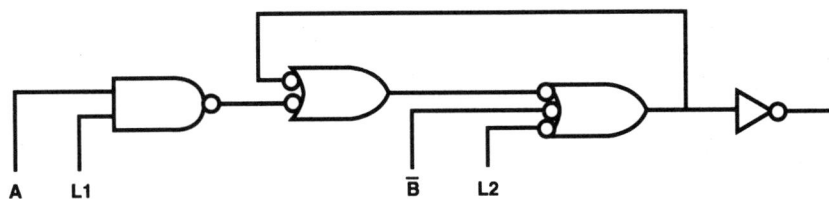

Figure 6-3. NAND sequential control circuit for *Example 6-2*.

Chapter Seven
CODES

A digital code is a binary representation of data. A binary code is a method for presenting data in a format suitable for digital circuits. A code can represent numbers, letters, and control signals. A useful machine code must be a binary code because digital circuits are on-off switching or binary circuits.

The decimal number system is not convenient for logic circuits because logic circuits operate on two-level signals. The binary number system is a more convenient system for logic circuits because it is represented by ones and zeros.

The decimal number system has 10 as its base or radix. The binary number system has 2 as its base or radix.

Example 7-1

Rewrite 123.45 as a polynomial with a base of 10:

$$123.45 = 1 \times 10^2 + 2 \times 10^1 + 3 \times 10^0 + 4 \times 10^{-1} + 5 \times 10^{-2}$$

Example 7-2

Rewrite 123.75 as a polynomial with a base of 2:

$$123.75 = 1 \times 2^6 + 1 \times 2^5 + 1 \times 2^4 + 0 \times 2^2 + 1 \times 2^1 + 1 \times 2^0 + 1 \times 2^{-1} + 1 \times 2^{-2}$$
$$(= 64 + 32 + 16 + 0 + 4 + 2 + 1 + 1/2 + 1/4)$$

The binary number system may be written more conveniently by representing each term of the polynomial by a 1 or a 0. A 1 indicates the presence of a term and a 0 represents the absence of a term. The polynomial of *Example 7-2* may be written as 1111011.11. The 0 indicates the absence of the 2^2 term in this polynomial.

OCTAL NUMBER SYSTEM

The octal number system uses 8 as its base.

Example 7-3

Write 49 in the octal number system:

Since 49/8 = 6 with a remainder of 1, 49_8 = 6 1.

Bits		Sum	Carry	Differ.	Borrow	Product
a	b	a + b		a - b		a.b
0	0	0	0	0	0	0
0	1	1	0	1	1	0
1	0	1	0	1	0	0
1	1	0	1	0	0	1

Table 7-1. Binary arithmetic operations.

BINARY ARITHMETIC

The basic arithmetic operations are shown in *Table 7-1*. As with decimal numbers, multiplication of binary numbers is performed by successive addition, while division is achieved by successive subtraction.

Example 7-4
```
1111     = carries of 1
1111.01 = 15.25₁₀
+
0111.10 = 7.50₁₀
10110.11 = 22.75₁₀
```

Example 7-5
```
1 = borrows of 1
10010.11 = 18.75₁₀
-
01100.10 = 12.50₁₀
00110.01 = 6.25₁₀
```

Example 7-6

```
    11001.1  = 25.5₁₀
  x   110.1  =  6.5₁₀
    ─────────
    000110011
    0000000
    0110011
    110011
    ─────────
  101000101.11 = 165.75₁₀
```

BINARY CODES

The binary number system has many advantages and is widely used in digital computers. To simplify the communication between men and machines, codes have been developed so that decimal digits are represented by a sequence of binary digits.

Decimal Digit	$2^3 = 8$	$2^2 = 4$	$2^1 = 2$	$2^0 = 1$
0	0	0	0	0
1	0	0	0	1
2	0	0	1	0
3	0	0	1	1
4	0	1	0	0
5	0	1	0	1
6	0	1	1	0
7	0	1	1	1
8	1	0	0	0
9	1	0	0	1

Table 7-2. 8421 code.

8421 CODE

The 8421 code is a weighted code because each binary digit is assigned a value. In each group of four bits or words, the sum of the weights of the bits that are 1 equals the decimal digit that they represent. The 8421 code is not self-complementary. The complement of a 1 is 0 and the complement of a 0 is 1. *Table 7-2* lists the 8421 code.

Example 7-7

$1101 = 8 + 4 + 0 + 1 = 13$

In the 8421 code, states or numbers 10-15 are illegal. When adding two 8421 code numbers, 0110 must be added to the raw sum if it is 10 or more.

Decimal Digit					Decimal Digit				
0	0	0	1	1	9	1	1	0	0
1	0	1	0	0	8	1	0	1	1
2	0	1	0	1	7	1	0	1	0
3	0	1	1	0	6	1	0	0	1
4	0	1	1	1	5	1	0	0	0

Complements

Table 7-3. Excess-3 code.

Example 7-8

0011 0110 1001
+
<u>0001 0111 1000</u>
0100 1110 0001 = raw sum
+
<u>0000 0110 0110</u>
0101 0100 0111

Decimal Digit	$2^1=2$	$2^2=4$	$2^1=2$	$2^0=1$	Decimal Digit	$2^1=2$	$2^2=4$	$2^1=2$	$2^0=1$
0	0	0	0	0	9	1	1	1	1
1	0	0	0	1	8	1	1	1	0
2	0	0	1	0	7	1	1	0	1
3	0	0	1	1	6	1	1	0	0
4	0	1	0	0	5	1	0	1	1

Complements

Table 7-4. 2421 code.

BCD CODE

Binary coded decimals, or BCD, represents specific decimal digits that are represented by a certain binary code. For example, in BCD code, the number 16 can be represented by 0001 for 1 and 0110 for 6.

EXCESS-3 CODE

The excess-3 code is an unweighted binary code. It is a self-complementing code. Each excess-3 code is formed by adding 0011 to each 8421 code word. The excess-3 code is listed in *Table 7-3*.

2421 CODE

The 2421 code is weighted and self-complementing. It is useful for digital-to-analog converters because it is a weighted code. *Table 7-4* lists the 2421 code.

GRAY CODE

The Gray code is a cyclical code because each successive bit differs by only one digit from the previous bit. This property is useful in analog-to-digital converters. *Table 7-5* lists the Gray code.

Decimal Number	Gray g_3	g_2	g_1	g_0	Binary b_3	b_2	b_1	b_0
0	0	0	0	0	0	0	0	0
1	0	0	0	1	0	0	0	1
2	0	0	1	1	0	0	1	0
3	0	0	1	0	0	0	1	1
4	0	1	1	0	0	1	0	0
5	0	1	1	1	0	1	0	1
6	0	1	0	1	0	1	1	0
7	0	1	0	0	0	1	1	1
8	1	1	0	0	1	0	0	0
9	1	1	0	1	1	0	0	1
10	1	1	1	1	1	0	1	0
11	1	1	1	0	1	0	1	1
12	1	0	1	0	1	1	0	0
13	1	0	1	1	1	1	0	1
14	1	0	0	1	1	1	1	0
15	1	0	0	0	1	1	1	1

Table 7-5. Gray code.

To convert from binary to Gray, the most significant bit (MSB) remains the same and each word is "exclusive ORed" with the next one. To convert from Gray to binary, MSB remains the same and each word is "exclusive ORed" with the next term. The result is "exclusive ORed" with the next, and so on.

Example 7-9

The 8421 code 0101 is converted to 0111 in the Gray code. The first term or MBS remains the same. "Exclusive ORing" 0 and 1 and "exclusive ORing" 1 and 0 results in a 1. Therefore, 0111 is the Gray equivalent of the 8421 code 0101.

Example 7-10

The 0111 Gray code can be converted back to the 8421 code of 0101. The MSB stays the same. The MSB of the 8421 code is "exclusive ORed" with the second bit of the Gray code to yield a 1. This 1 is "exclusive ORed" with the third bit of the Gray code to yield a 0. This 0 is "exclusive ORed" with the fourth bit of the Gray code to yield a 1. Therefore, the Gray code 0111 is converted to the 8421 code 0101.

ERROR DETECTION AND CORRECTION

Transmission errors can occur due to a noisy transmission line or equipment failure. A redundant or extra bit may be added to a data block for checking; this is known as *parity*. The number of ones in a data block are counted. If the number of ones is odd, then the parity bit is 0 for odd parity. If the number of ones is even, then the parity bit is 1 for odd parity. Odd parity is more common than even parity because it can distinguish between zeros and "not ones".

Decimal Digit	5	0	4	3	2	1	0
0	0	1	0	0	0	0	1
1	0	1	0	0	0	1	0
2	0	1	0	0	1	0	0
3	0	1	0	1	0	0	0
4	0	1	1	0	0	0	0
5	1	0	0	0	0	0	1
6	1	0	0	0	0	1	0
7	1	0	0	0	1	0	0
8	1	0	0	1	0	0	0
9	1	0	1	0	0	0	0

Table 7-6. Biquinary code.

BIQUINARY CODE

The biquinary code is a seven-bit weighted decimal code. It has a two-bit and a five-bit field. It always has two ones, and therefore never needs a parity check. *Table 7-6* lists the biquinary code.

ASCII CODE

The American Standard Code for Information Interchange is used to represent alpha-numeric characters. It is a seven-bit code with a three-bit and a four-bit field.

HAMMING CODE

The Hamming code is a seven-bit code. The first two positions are parity checking bits, and the fourth position is also a parity checking bit. *Table 7-7* lists the Hamming code for the 8421 code.

Decimal Digit	Position	1 P_1	2 P_2	3 m_1	4 P_3	5 m_2	6 m_3	7 m_4
0		0	0	0	0	0	0	0
1		1	1	0	1	0	0	1
2		0	1	0	1	0	1	0
3		1	0	0	0	0	1	1
4		1	0	0	1	1	0	0
5		0	1	0	0	1	0	1
6		1	1	0	0	1	1	0
7		0	0	0	1	1	1	1
8		1	1	1	0	0	0	0
9		0	0	1	1	0	0	1

Table 7-7. Hamming code for the 8421 code.

PROBLEMS

Problem 7-1.

Rewrite 101.375 as a polynomial with a base of 10.

Problem 7-2.

Rewrite 101.375 as a polynomial with a base of 2.

Problem 7-3.

Rewrite the solution of *Problem 7-2* as a binary word.

Problem 7-4.

Write the octal number word for 59.

Problem 7-5.

Solve: 1000100110/11001.

Problem 7-6.

Write the BCD code if the BCD code is the 8421 code where numbers 10-15 are legal.

Problem 7-7.

Convert the binary word 1101 into a Gray word.

Problem 7-8.

Convert the Gray word 1011 into a binary word.

Chapter Eight
REGISTERS

A register is an assembly of flip-flops capable of storing binary data. Data may be transferred in or out of the register in parallel or in serial.

PARALLEL ENTRY

Parallel entry is when all of the bits of a binary word are transferred at the same time. When two people walk arm-in-arm through a door, they are entering in parallel. Parallel entry is either a one-step or two-step process. Each step may be either asynchronous or synchronous. The two-step operation is:

> (a) Clear all flip-flops with a common CLEAR signal serving all flip-flops.
> (b) Set only the flip-flops for which the input data is 1. This step is controlled by a common ENTER or TRANSFER signal.

The reverse process is also possible:

> (a) Preset all flip-flops with a common PRESET signal.
> (b) Clear only the flip-flops for which the input data is 0.

Both steps in each alternative may be synchronous. Also, both may be asynchronous, or one may be synchronous and the other (usually the presetting step) asynchronous.

A one-step or JAM operation omits the initial common preclearing or presetting stage. Data ones and zeros are entered simultaneously by setting or clearing the appropriate flip-flop. This step is initiated by a common control signal and may be synchronous or asynchronous. Unless the clock is gated, a control gate is needed on each logic input of each flip-flop. The one-step process uses more hardware, but takes less time than the two-step process.

TIMING

Synchronous operation is enabled by an ENABLE signal and is timed by the system clock. All flip-flop outputs change state on a specified edge of the clock pulse. A timing diagram for synchronous operation is shown in *Figure 8-1*.

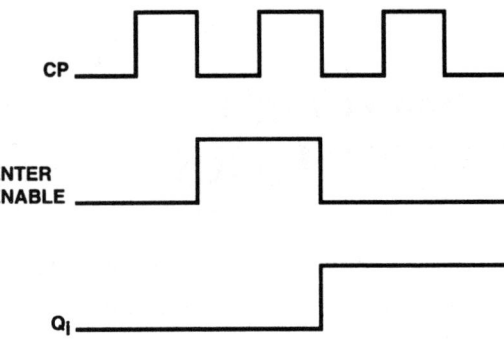

Figure 8-1. Timing diagram for synchronous operation.

Asynchronous operation of edge-triggered flip-flops uses direct SET and CLEAR inputs, and the entry is timed by the ENTER control signal. It is common to see asynchronous operations inserted between clock pulses. This reduces the number of clock periods required to complete a digital process. Sufficient time must be available between clock pulses for the flip-flop outputs to settle after a clocked operation and after an asynchronous data entry. The timing diagram of an asynchronous operation is shown in *Figure 8-2*.

Example 8-1

The timing diagram of a two-step operation consisting of synchronous clearing and entering of ones is shown in *Figure 8-3*.

Example 8-2

The timing diagram of a two-step operation consisting of asynchronous preclearing and synchronous entering of ones is shown in *Figure 8-4*. The whole entry process occurs during one clock period.

Example 8-3

The timing diagram of a two-step process consisting of asynchronous preseting and entering of zeros is shown in *Figure 8-5*.

Example 8-4

The timing diagram for a synchronous jam entry is shown in *Figure 8-6*.

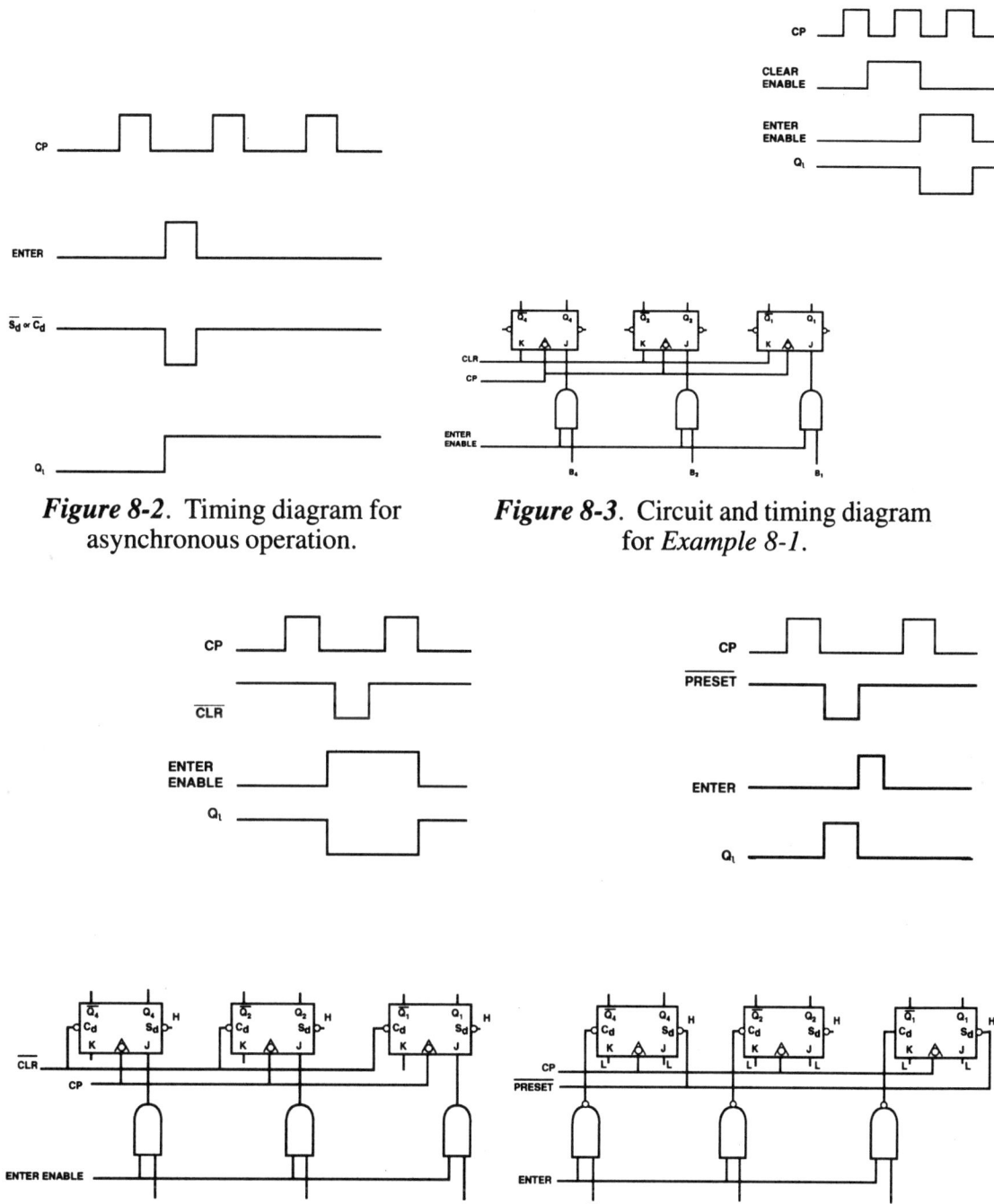

Figure 8-2. Timing diagram for asynchronous operation.

Figure 8-3. Circuit and timing diagram for *Example 8-1*.

Figure 8-4. Circuit and timing diagram for *Example 8-2*.

Figure 8-5. Circuit and timing diagram for *Example 8-3*.

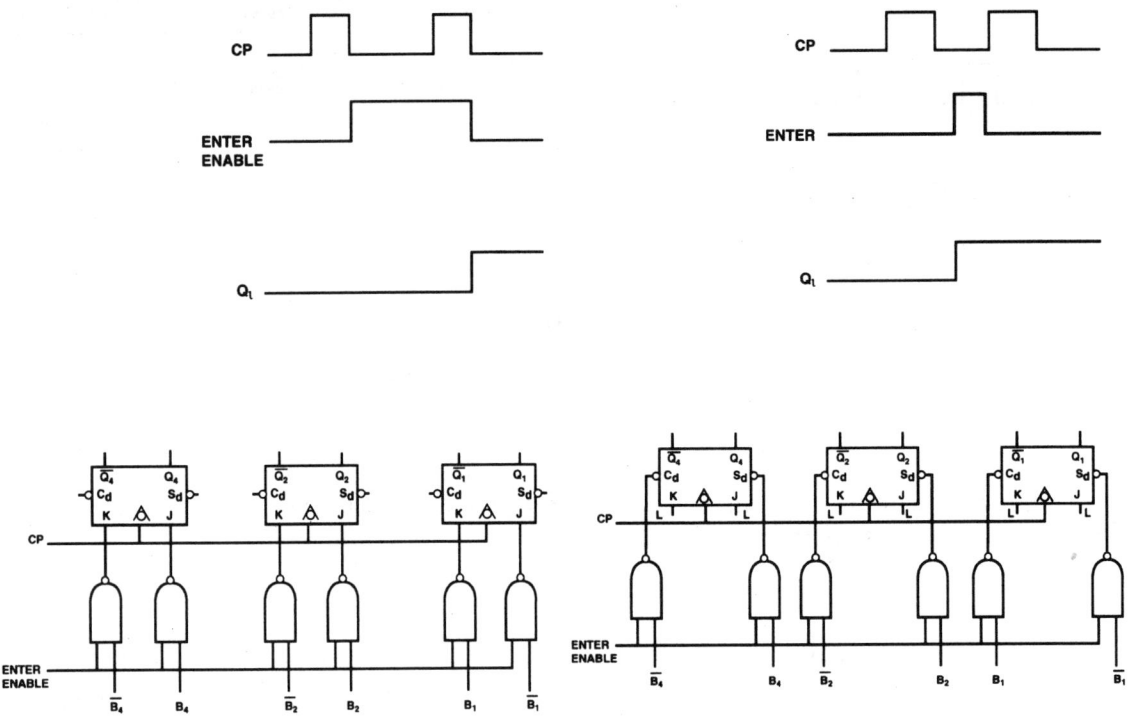

Figure 8-6. Circuit and timing diagram for *Example 8-4*.

Figure 8-7. Circuit and timing diagram for *Example 8-5*.

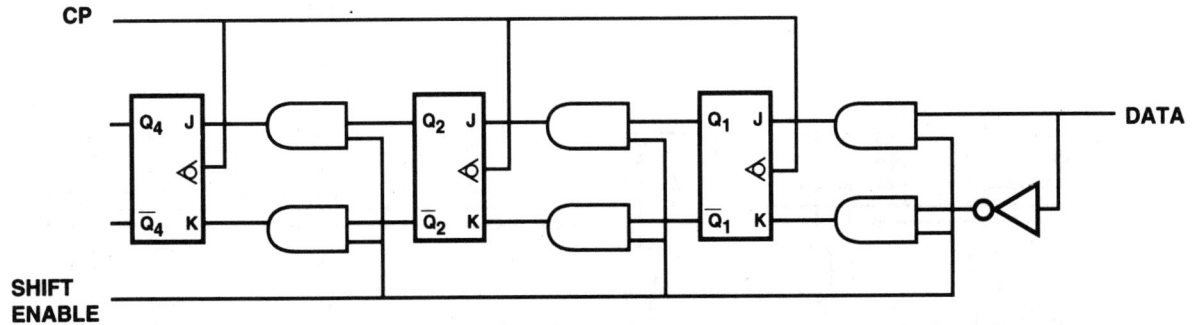

Figure 8-8. Controlled shift register.

Example 8-5

The timing diagram for an asynchronous jam transfer is shown in *Figure 8-7*.

SERIAL ENTRY

Serial entry of binary data is the entry of the data in or out of a shift register one bit at a time. When people enter through a door in single file, they are entering serially. The register must be made of either master-slave or edge-triggered flip-flops so that a data bit at the input will not race through the register when the clock pulse is high. The shifting process may be controlled by gating the clock or by coupling the output of a flip-flop to the input of the next one with AND gates. When the SHIFT control is LOW, the AND gates are disabled and no shifting can take place. The shift register is unaffected by clock pulses because all SET and CLEAR inputs are LOW. A controlled shift register is shown in *Figure 8-8*. Shift registers are often used in arithmetic circuits. A shift register is also used to generate a continuously repeating binary pattern which may be shown by a state diagram. Data may be shifted to the right or to the left. *Figure 8-9* shows a shift left/shift right register.

SHL	SHR	Operation
0	0	None
0	1	Shift Right
1	0	Shift Left
1	1	Not Allowed

COUNTERS

A counter is a circuit with one input and one output. The input terminal receives pulse signals. The output starts at some initial state and counts through a specific sequence until it reaches its initial state again. Then the sequence is repeated. Each step of the sequence is called a *state*. A binary counter requires 2^n flip-flops such that $2^n = X$ where X is the number of states in the counter's sequence. A counter with an eight-state sequence requires three flip-flops.

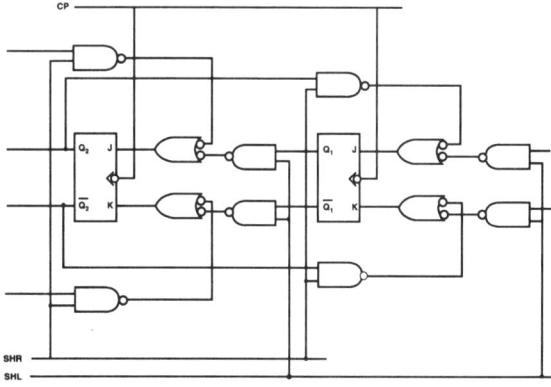

Figure 8-9. Shift left/shift right register.

RING COUNTER

The ring counter is a counter with the output of a flip-flop connected to the input of the next flip-flop. *Figure 8-10* shows a schematic of a ring counter. When the circuit is initialized, $Q_0 = 1$ and all the other outputs are 0. Clocking shifts the 1 around the closed loop.

COMPLEMENTING RING COUNTER

The complementing ring counter has the Q output of the last stage connected to the CLEAR input of the first stage, as shown in *Figure 8-11*. After initialization, the register fills with ones and then with zeros. The sequence of states is also shown in *Figure 8-11*. The sequence of states includes four states. What happened to the other four states? If the counter is initialized at one of the other four states, the sequence(s) obtained accounts for the remaining four states and is also cyclic. This is called the secondary mode sequence(s), and it is also shown in *Figure 8-11*. The Johnson counter, or shift counter, always has a sequence with an even number of states.

A ring counter can be self-starting if Q output(s) is connected to the CLEAR input with an AND gate or by connecting Q output(s) to the CLEAR input with a NAND gate.

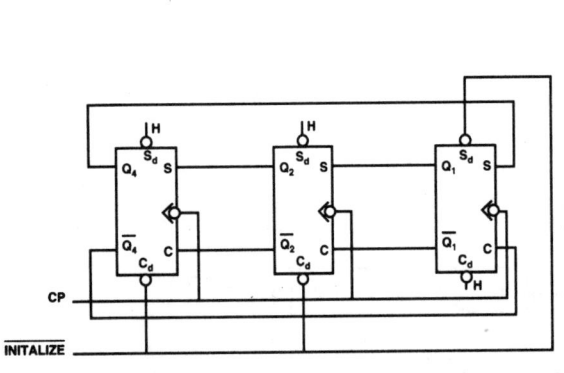

Figure 8-10. Ring counter.

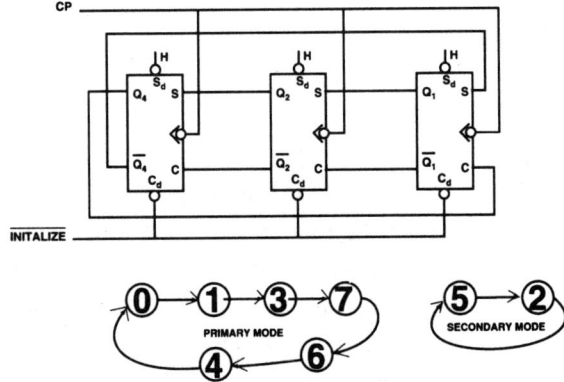

Figure 8-11. Complementing ring counter.

DIMINISHED COMPLEMENTING RING COUNTER

The diminished complementing ring counter has a sequence with an odd number of states. This is achieved by either bypassing the 0000 state or the 1111 state. The circuit and state diagram of a diminished complementing ring counter are shown in *Figure 8-12*.

PROBLEMS

Problem 8-1.
What is a register?

Problem 8-2.
What is parallel entry?

Problem 8-3.
What is the advantage of asynchronous operation? Why?

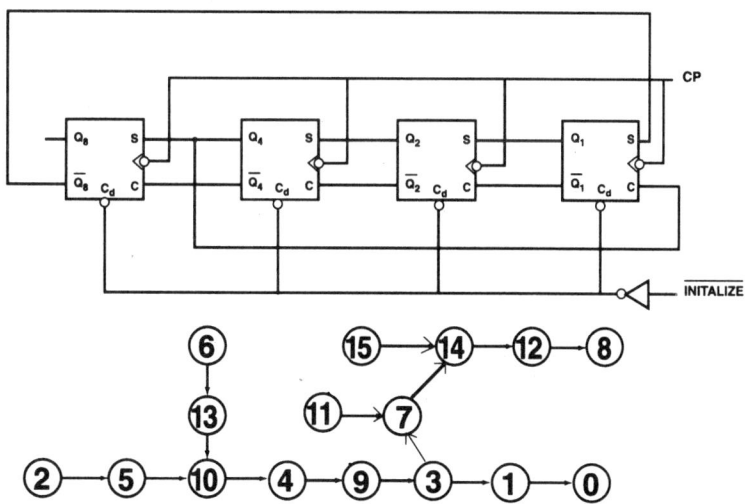

Figure 8-12. Diminished complementing ring counter.

Problem 8-4.
Draw the timing diagram for entry into a gated latch register.

Problem 8-5.
In serial data entry, why must the flip-flops be master-slave or edge-triggered flip-flops?

Problem 8-6.
Draw a self-starting counter.

Problem 8-7.
Draw a self-starting complementing ring counter.

Problem 8-8.
Design a binary counter and draw its state diagram.

Chapter Nine
ENCODERS, DECODERS & MULTIPLEXERS

Encoders change many lines to a few lines, and *decoders* change a few lines to many lines. *Multiplexers* are data selectors.

ENCODERS

The keyboard encoder translates the signal produced by each key action into a binary number which is stored in a register. The register is called a *keyboard buffer register*. *Figure 9-1* is the circuit of a decimal-digit-to-BCD character converter.

DECODERS

The decoder produces a separate output for each possible input. There are several types of decoders.

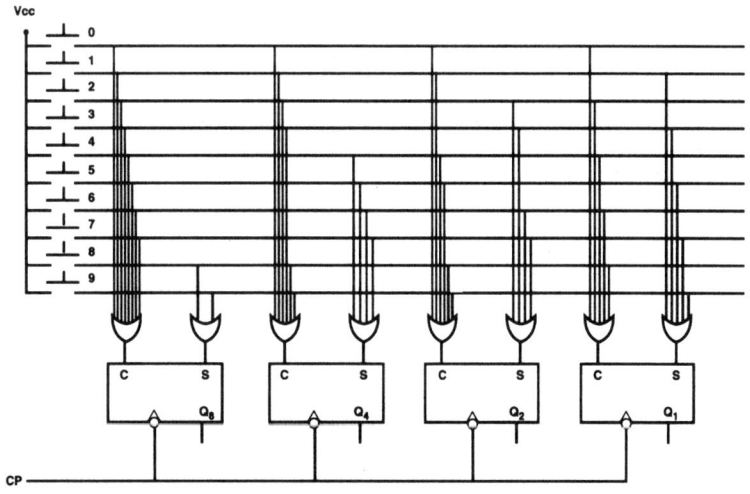

Figure 9-1. Decimal digit-to-BCD character converter.

Rectangular Decoder

The rectangular decoder can have only one AND output high at any time, and is labeled with the state present at that time. If NAND gates are used, the selected output is low. When NOR gates are used, the connections to the two outputs of each flip-flop must be interchanged. *Figure 9-2* shows a circuit for a rectangular decoder. To decode a three flip-flop register, eight three-input AND gates are required. A decoder for a register of n flip-flops requires 2^n, n-input AND gates. If n is large, a lot of gates and input connections are required. MSI (medium scale integration) components can be used to avoid the use of many gates.

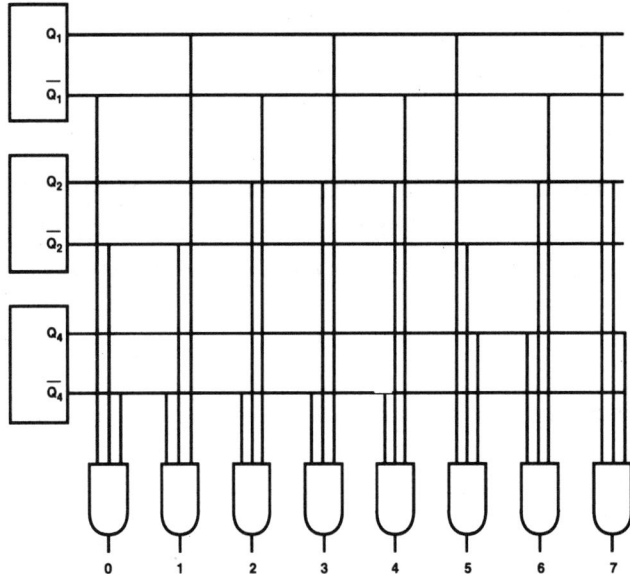

Figure 9-2. Rectangular decoder.

Tree Decoder

In a tree decoder, two flip-flops may be decoded using four two-input AND gates. Each output is AND gated, with the next input to produce eight decoded outputs. This process is repeated over again to yield sixteen decoded outputs. *Figure 9-3* shows a tree decoder. A tree decoder with a four-bit register requires 28 AND functions and 56 inputs. The rectangular decoder needs 16 AND functions and 64 inputs. The tree decoder has fewer input interconnections.

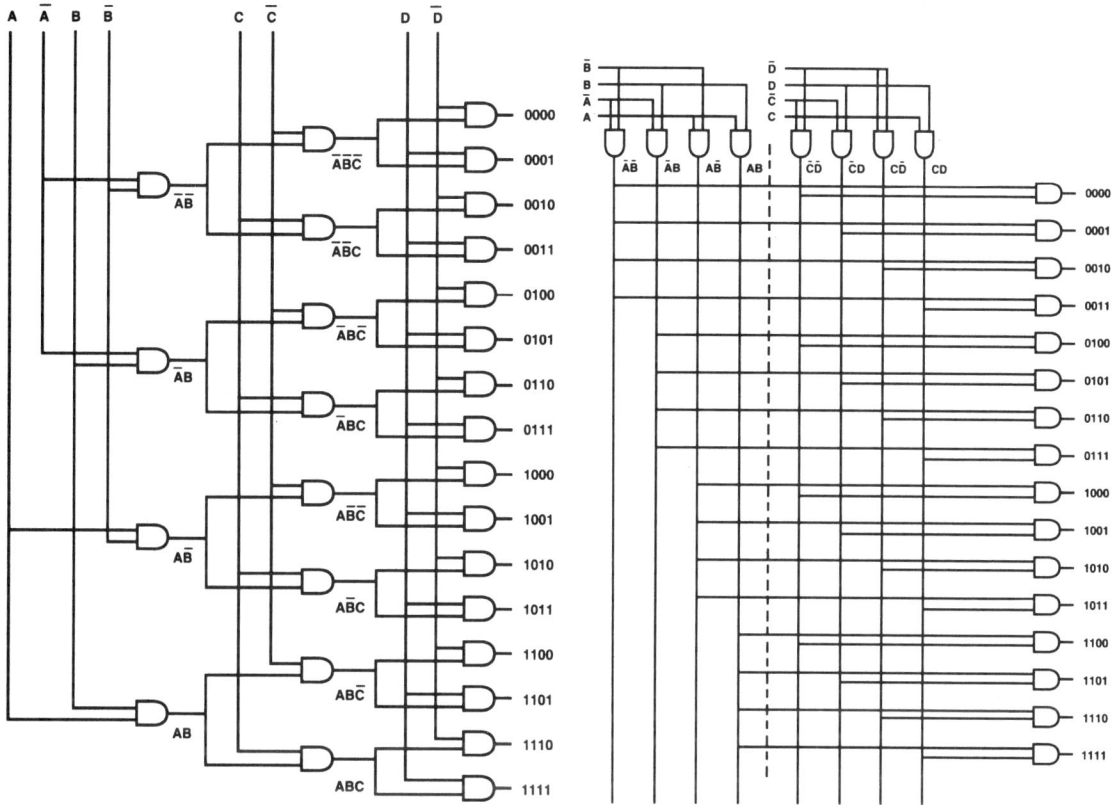

Figure 9-3. Tree decoder.   *Figure 9-4.* Dual tree decoder.

Dual Tree Decoder

In a dual tree decoder, the register is divided into two halves. If each half contains more than two or three flip-flops, they are again divided into two halves. The process is repeated until further division is impossible. Each group of flip-flops is decoded, and each decoded output of one group is AND gated with each decoded output of all of the other groups, using a tree structure if necessary to provide the final decoder outputs. *Figure 9-4* shows a dual tree decoder. This type of decoder provides savings in hardware and input connections. The trade-off is speed because several levels of delay in the dual tree decoder increase the propagation delay of the circuit.

To decode a twelve flip-flop register, a rectangular decoder requires 4096 twelve-input AND gates, totaling 49152 inputs. The tree decoder needs 8188 AND gates totaling 16376 inputs. The dual tree decoder requires 8544 inputs consisting of 32 three-input AND gates and 4224 two-input AND gates.

1/n Decoder

MSI decoders are available as 1/4 (one out of four), 1/8 and 1/16 devices. *Figure 9-5* shows a 1/4 decoder. It has two address or register inputs, an active low-enable input, and four active low outputs. It can decode a two flip-flop register by connecting $E = 0$, $B = Q2$, and $A = Q1$. The truth table for this circuit is in *Table 9-1*.

1/2n Decoder

A 1/2n decoder may be designed using two 1/n decoders. The circuit is shown in *Figure 9-6*. A 1/8 decoder can be designed using two 1/4 decoder plus an inverter, as shown in *Figure 9-6*. When Q4 is low, the right-hand module is selected and the output corresponding to the present state of Q1 and Q2 is low. All other outputs are high. If Q4 is high, the left-hand module is selected and one of its outputs will be low and all other outputs high.

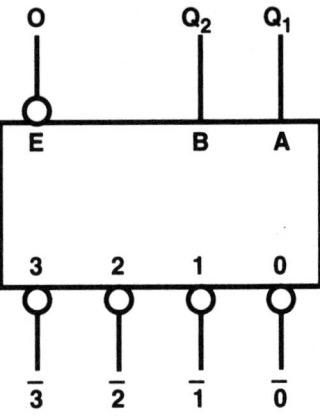

Figure 9-5. 1/4 decoder.

1/n² Decoder

A 1/16 decoder can be designed using five 1/4 decoders, as shown in *Figure 9-7*. The first module is enabled all the time by a low on its E input. This decodes the two most significant bits, Q8 and Q4. Each of these four variables enables one of the four remaining modules.

Modules in other families may have active high outputs. These modules also use an active high enable. The selected output of one module may be used to enable another module.

1/10 Decoder

The 1/10 decoder is called a *decade decoder*. *Figure 9-8* shows a 1/20 decoder designed using two 1/10 decoders and an inverter. The truth table for the 1/20 decoder is also shown in *Figure 9-8*. The 1/10 decoder is available as an integrated circuit.

E	B	A	3	2	1	0
0	0	0	1	1	1	0
0	0	1	1	1	0	1
0	1	0	1	0	1	1
0	1	1	0	1	1	1
1	0	0	1	1	1	1
1	0	1	1	1	1	1
1	1	0	1	1	1	1
1	1	1	1	1	1	1

Table 9-1. Truth table for 1/4 decoder.

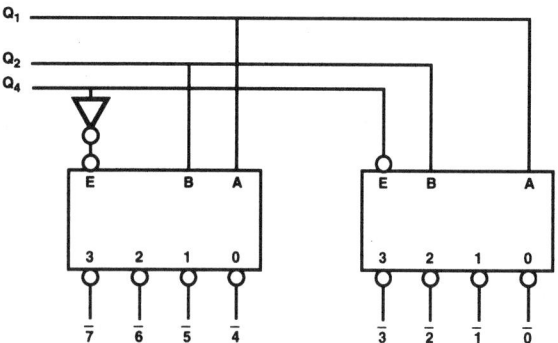

Figure 9-6. 1/8 decoder.

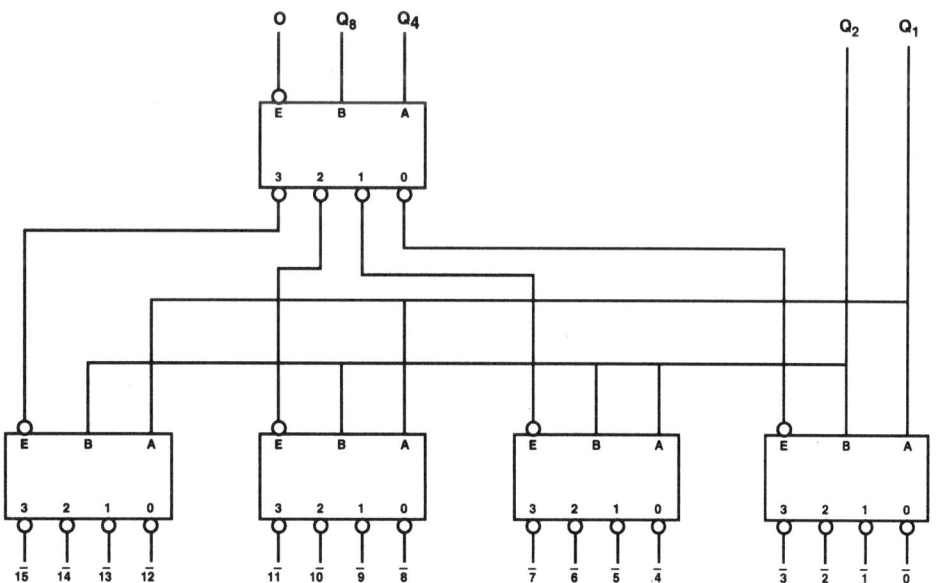

Figure 9-7. 1/16 decoder.

62 / Digital Electronics

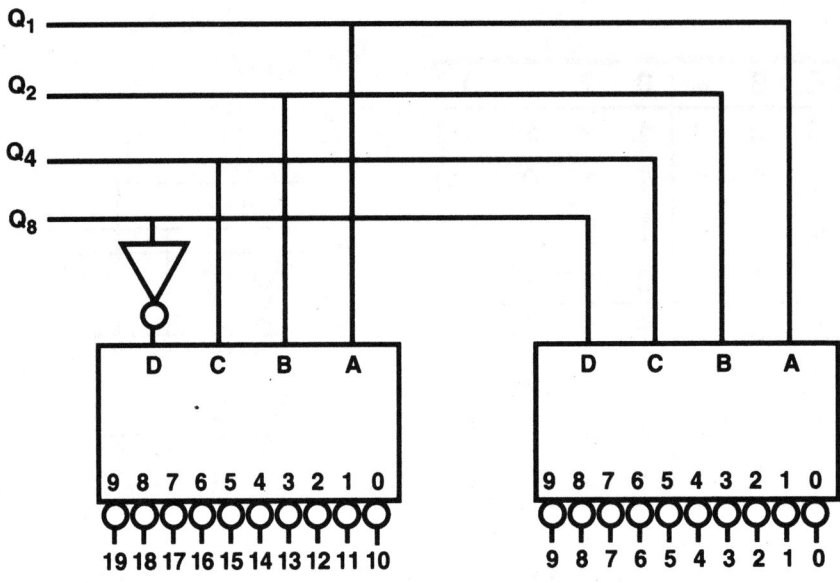

Figure 9-8. 1/20 decoder.

Input Code				Selected Outputs		Most Useful Outputs
				Protected Version	Minimized Version	
0	0	0	0	0, 18	0, 18	0
0	0	0	1	1, 19	1, 19	1
0	0	1	0	2, -	2, 18	2
0	0	1	1	3, -	3, 19	3
0	1	0	0	4, -	4, 18	4
0	1	0	1	5, -	5, 19	5
0	1	1	0	6, -	6, 18	6
0	1	1	1	7, -	7, 19	7
1	0	0	0	8, 10	8, 10	10
1	0	0	1	9, 11	9, 11	11
1	0	1	0	-, 12	8, 12	12
1	0	1	1	-, 13	9, 13	13
1	1	0	0	-, 14	8, 14	14
1	1	0	1	-, 15	9, 15	15
1	1	1	0	-, 16	8, 16	16
1	1	1	1	-, 17	9, 17	17

MULTIPLEXERS

The multiplexer is a data selector. The multiplexer has *n* addresses, which decode and enable the connection of any one of 2^n data inputs to a common data output. Most multiplexers have an enable input but a few multiplexers have complementary outputs.

The two-line-to-one-line multiplexer is shown in *Figure 9-9*. It connects either of the two inputs, I0 or I1, to the output, Z, depending on the state of the select line. S.E is a chip enable input.

A four-line-to-one-line multiplexer is shown in *Figure 9-10*. Two selection bits, S1 and S2, are needed to select one of four lines.

Eight-line-to-one-line and sixteen-line-to-one-line multiplexers are shown in *Figure 9-11*.

APPLICATIONS FOR MULTIPLEXERS

Two 8-to-1 multiplexers may be used to design a 16-to-1 multiplexer, as shown in *Figure 9-12*. The most significant address bit is used to enable one side or the other.

A multiplexer may be used as a parallel-to-serial converter, as shown in *Figure 9-13*. The two-bit counter generates the selection bits. The input data register outputs the change state on the trailing edge of every fourth clock pulse.

A data router can be designed using a quad (four in each integrated circuit) two-to-one multiplexer to connect registers A-to-C, when select line SEL = 0 and registers B-to-C when SEL = 1. The circuit for the data router is shown in *Figure 9-14*.

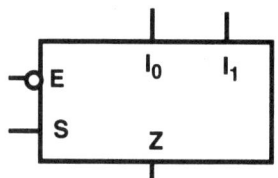

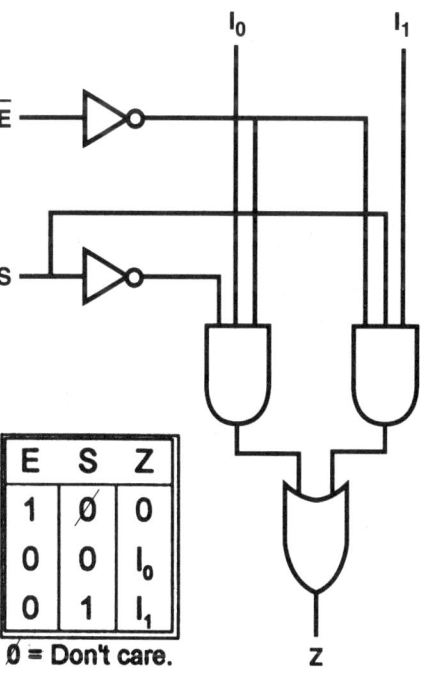

E	S	Z
1	∅	0
0	0	I_0
0	1	I_1

∅ = Don't care.

Figure 9-9. Two-line to one-line multiplexer.

DEMULTIPLEXERS

The demultiplexer takes the data of a single input and distributes it to 2^n data lines, as specified by an *n*-bit address input. The demultiplexing function can be achieved using an MSI decoder which has an enable input. The selected output follows the E input. The E is used as the data input while the decoder outputs are used as the distributed data outputs. Output data is available on each line only during the time slot defined by the inputs Q2 and Q1. A demultiplexer is shown in *Figure 9-15*.

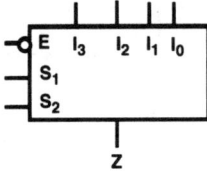

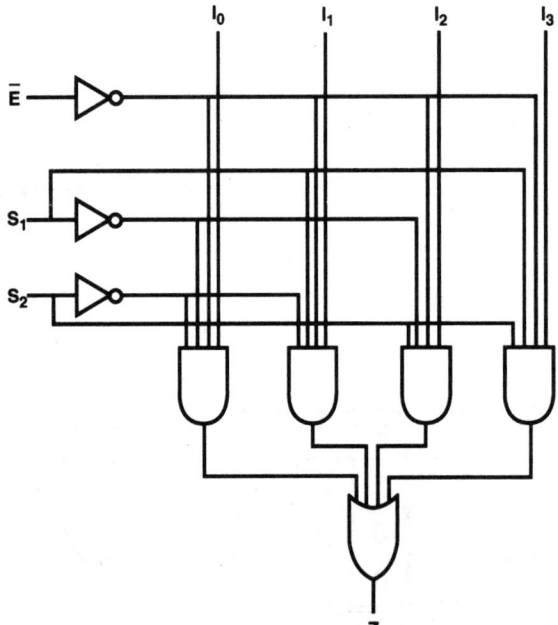

E	S1	S2	Z
0	0	0	I_0
0	0	1	I_1
0	1	0	I_2
0	1	1	I_3
1	0	0	0
1	0	1	0
1	1	0	0
1	1	1	0

Figure 9-10. Four-line to one-line multiplexer.

Encoders, Decoders & Multiplexers / 65

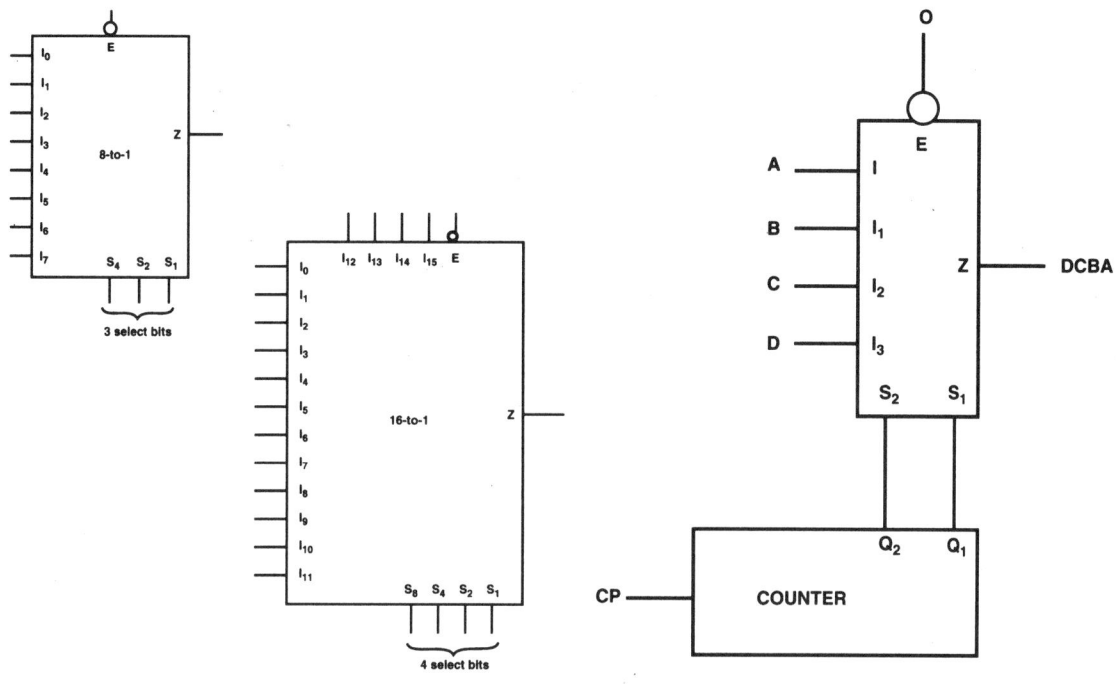

Figure 9-11. Eight-line to one-line and sixteen-line to one-line multiplexers.

Figure 9-13. Parallel-to-series converter.

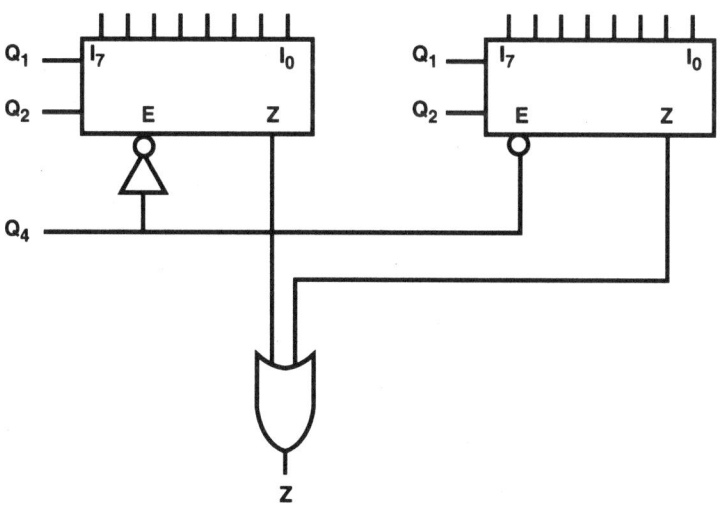

Figure 9-12. 16-to-1 multiplexer.

PROBLEMS

Problem 9-1.

Define encoder, decoder, and multiplexer.

Problem 9-2.

Compare rectangular, tree, and dual tree decoders.

Problem 9-3.

Redraw the circuit of *Figure 9-2* using NOR gates instead of AND gates.

Problem 9-4.

Draw a rectangular decoder using diodes in place of AND gates.

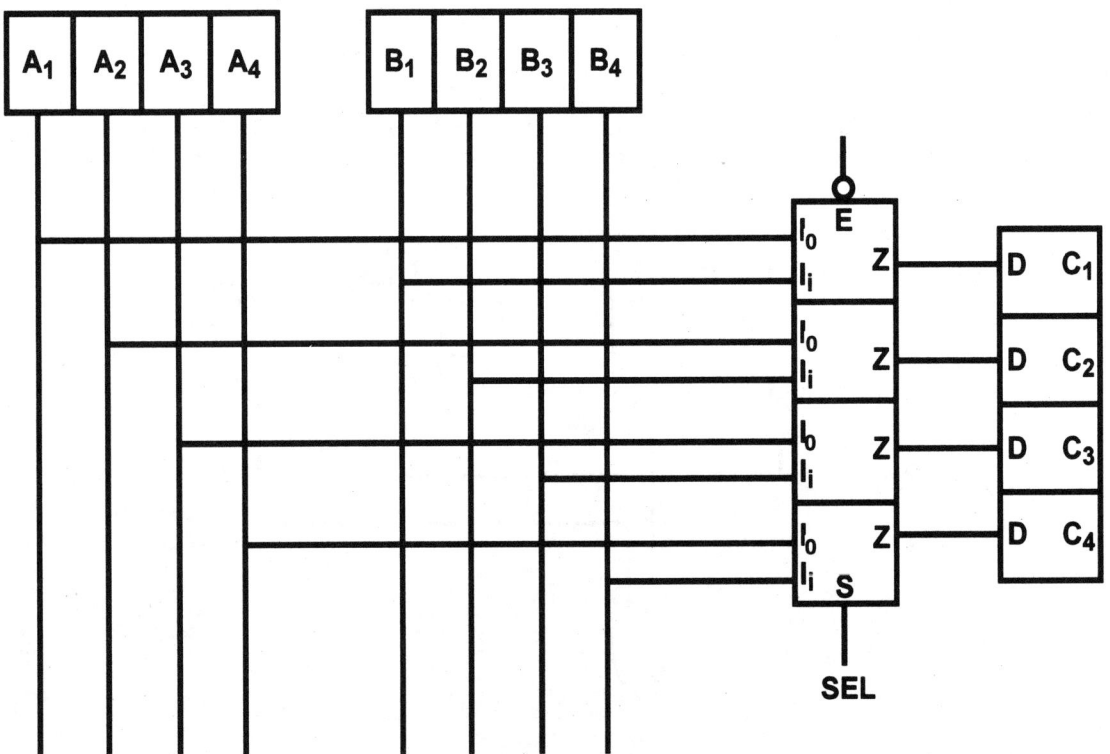

Figure 9-14. Data router.

Problem 9-5.

How many AND gates with how many inputs are required to decode five flip-flops?

Problem 9-6.

Draw a 1/8 decoder module.

Problem 9-7.

Design a 1/16 decoder using 1/8 decoders.

Problem 9-8.

Design a 1/64 decoder using 1/8 decoders.

Problem 9-9.

Design a 1/100 decoder using decade (1/10) decoders.

Problem 9-10.

Design a register mode control using multiplexers.

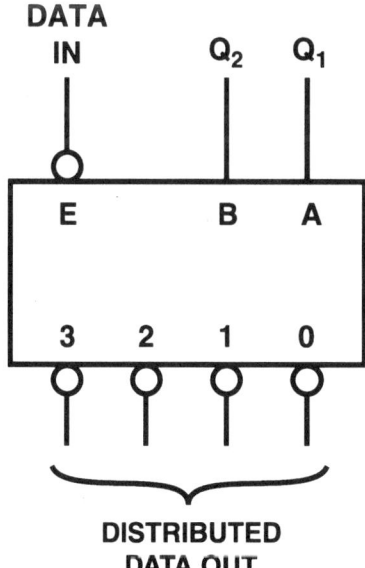

E	B	A	3	2	1	0
0	0	0	1	1	1	0
0	0	1	1	1	0	1
0	1	0	1	0	1	1
0	1	1	0	1	1	1
1	0	0	1	1	1	1
1	0	1	1	1	1	1
1	1	0	1	1	1	1
1	1	1	1	1	1	1

Figure 9-15. Demultiplexer.

Chapter Ten
COMPARATOR & EXCLUSIVE-OR CIRCUITS

The exclusive-OR (XOR) circuit provides a 1-level output when the two inputs, A and B, are *different*. The comparator provides a 1-level output when the two inputs, A and B, are the *same*. The circuits and truth tables for the XOR and comparator circuits are shown in *Figure 10-1*.

The XOR gate has the following properties:

(a) Commutativity: $A \oplus B = B + A$
(b) Associativity: $(A \oplus B) \oplus C = A \oplus (B \oplus C) = A \oplus B \oplus C$
(c) Distributivity: $(AB) \oplus (AC) = A(B \oplus C)$

If inverted inputs are not available, the exclusive-OR function can be achieved by the circuit shown in *Figure 10-2*. The truth tables and Karnaugh maps for 2-input XOR and comparator circuits are shown in *Table 10-1*.

A	B	$f_\oplus$
0	0	0
0	1	1
1	0	1
1	1	0

A	B	f_c
0	0	1
0	1	0
1	0	0
1	1	1

```
      0  1  A
   0 [0][1]   f = A⊕B
   1 [1][0]
   B
```

```
      0  1  A
   0 [1][0]   f = A⊕B
   1 [0][1]
   B
```

Table 10-1. Truth tables and Karnaugh maps for XOR and comparator circuits.

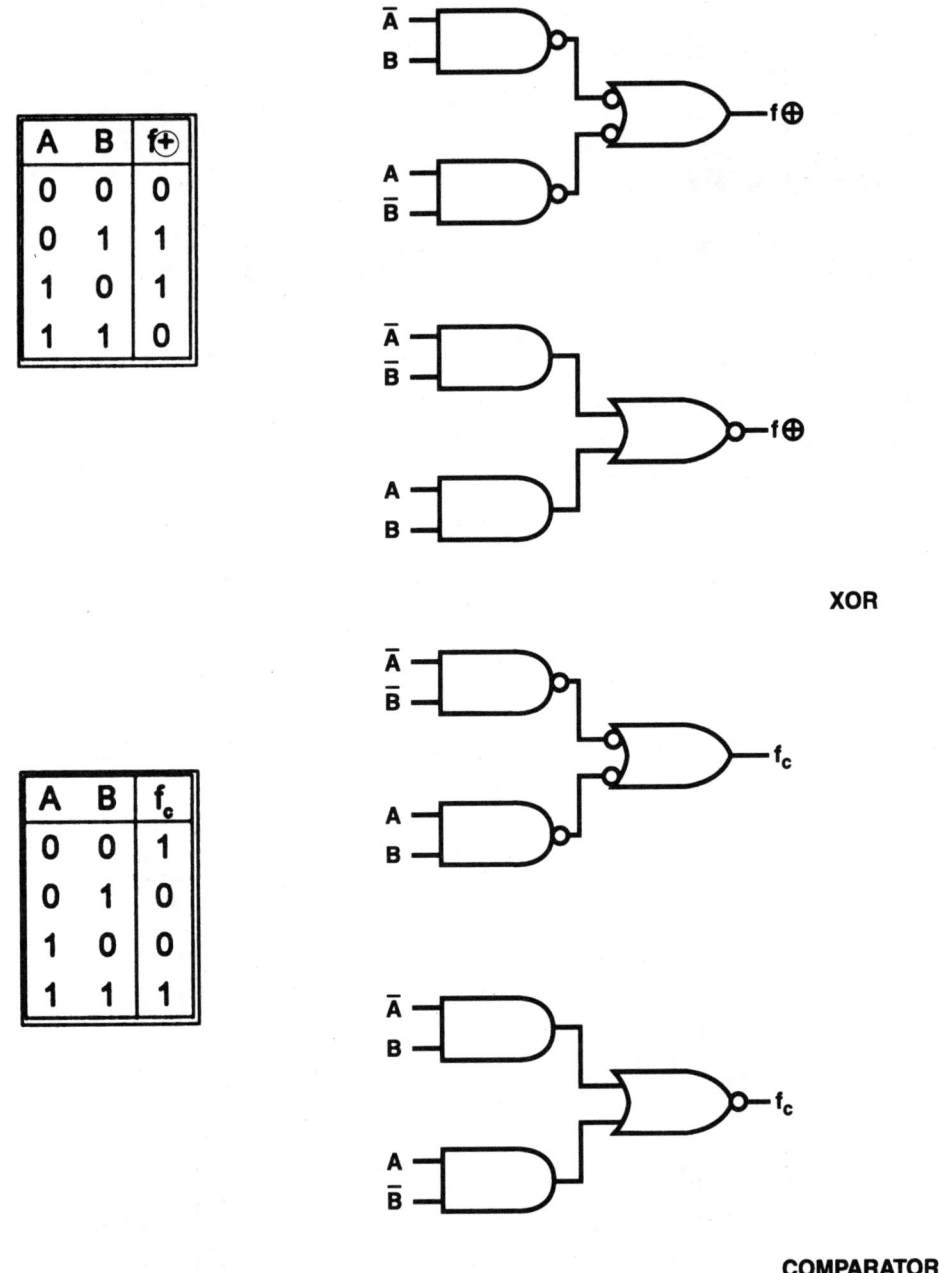

Figure 10-1. XOR and comparator circuits.

APPLICATIONS

An XOR gate can be used as a controlled inverter. When control line C = 0, Z = Y and Z = $\overline{Y}$ when C = 1. The circuit is shown in *Figure 10-3*.

A parity check circuit can be designed with XOR gates, as shown in *Figure 10-4*. If there are an odd number of ones on the five inputs A, B, C, D and P, then f = 1.

A Gray-code-to-binary-code converter is shown in *Figure 10-5*. The circuit operation can be verified by inputting any Gray code. The output should be the equivalent binary code.

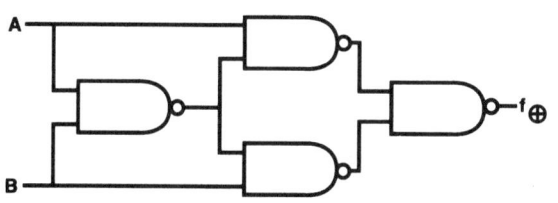

Figure 10-2. XOR circuit when inverted inputs are not available.

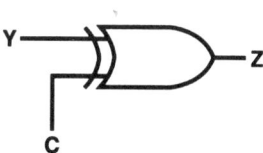

Figure 10-3. Controlled inverter.

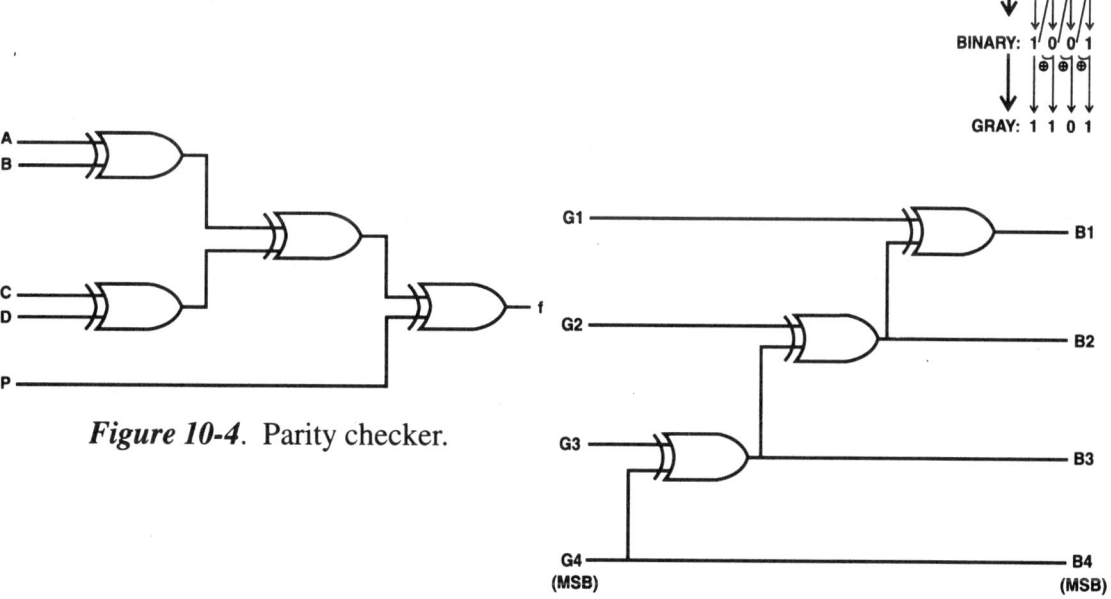

Figure 10-4. Parity checker.

Figure 10-5. Gray code to binary code converter.

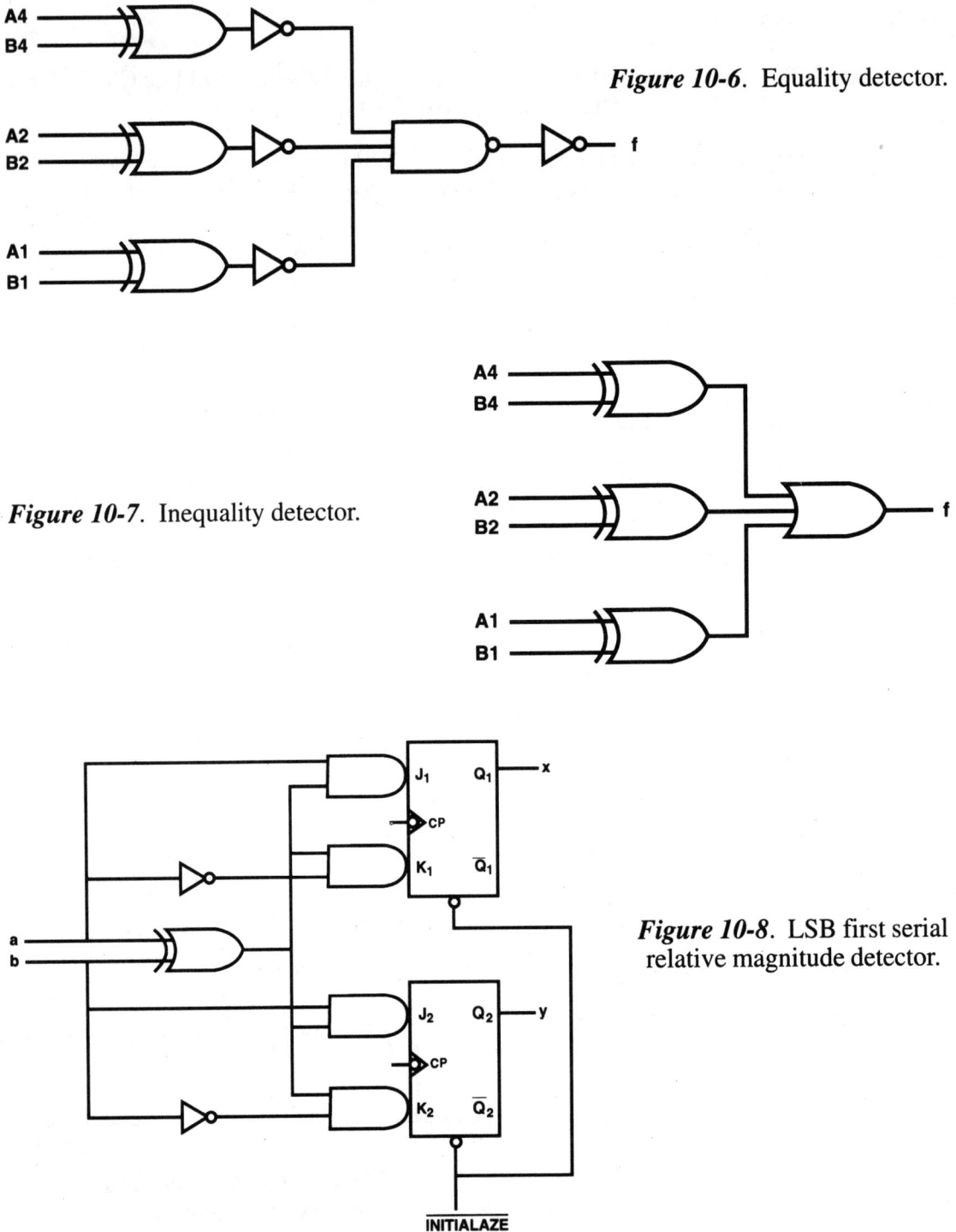

Figure 10-6. Equality detector.

Figure 10-7. Inequality detector.

Figure 10-8. LSB first serial relative magnitude detector.

COMPARATOR CIRCUITS

An *n*-bit comparator is a circuit which compares the magnitude of two numbers, X and Y. A comparator has three outputs: f1, f2 and f3 such that f1 = 1 if and only if X > Y; f2 = 1 if and only if X = Y; f3 = 1 if and only if X < Y. 4-bit comparators are available as integrated circuits.

An equality detector outputs a 1 when its inputs are equal. *Figure 10-6* shows an equality detector circuit. An inequality detector outputs a 1 when its inputs are different. An inequality detector circuit is shown in *Figure 10-7*.

Serial data is often transmitted LSB (least significant bit) first because this is more convenient for arithmetic operations. *Figure 10-8* shows a circuit for an LSB first-serial relative magnitude detector. An inequality detected in any bit position propagates to the output only if all higher order bits are equal. Any inequality overrides all previous decisions:

If a = b, XOR output = 0, J1 = K1 = J2 = K2 = 0; X and Y are unchanged.
If a = 1 and b = 0, J1 = 1, K1 = 0, J2 = 0 and K2 = 1; X becomes 1 and Y becomes 0.
Similarly, if a = 0 and b = 1, X becomes 0 and Y becomes 1.

The truth table (after the MSB is clocked in) for the serial magnitude detector is also shown in *Figure 10-8*.

There are also MSB first-relative magnitude detectors. If the most significant bit of the two words being compared are not equal, the circuit outputs a 1 to indicate that either A > B or A < B. If the MSBs of the two words are equal, the circuit outputs a 0 to indicate that A = B. If this is the case, the next most significant bits of the two words are compared and the process is repeated. A bit position is tested only if all higher order bit pairs are equal. When an inequality is found, all lower order bits are prevented from influencing the result. The circuit of an MSB first-relative magnitude detector is shown in *Figure 10-9*.

PROBLEMS

Problem 10-1.
Draw the circuits and Karnaugh maps for 4-input XOR and comparator circuits.

Problem 10-2.

Draw a parity-check circuit to check the parity of a 4-bit character.

Problem 10-3.

Design a binary-code-to-Gray-code converter that uses the algorithm shown in *Figure 10-5*.

Problem 10-4.

Draw an LSB first-relative magnitude detector using 4-to-1 multiplexers.

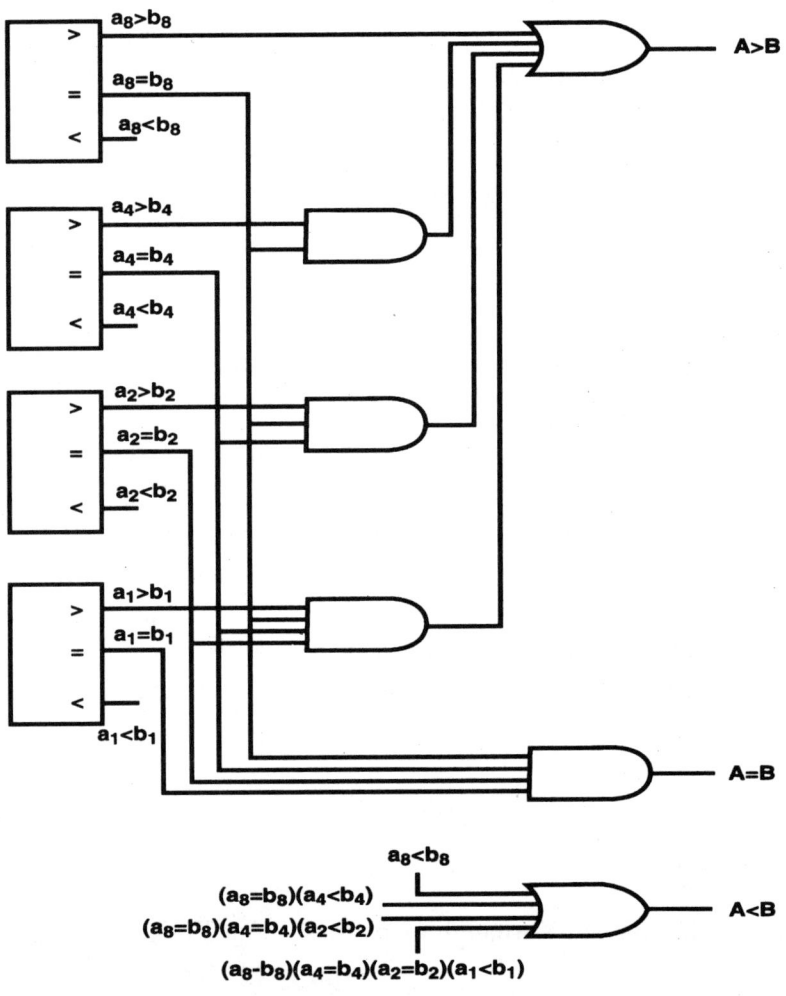

Figure 10-9. MSB first relative magnitude detector.

Chapter Eleven
COUNTERS

Counters have many applications in digital systems. They are often used as event counters, for example, to count the number of automobiles that cross an intersection each hour. Counters are also used for the timing of digital systems.

The simplest counters count in a binary fashion, and they repeat their count sequence every 2^n input pulses, where n is an integer equal to the number of flip-flops used to construct the counter. Other scales may be used in a counter; the decade counter is a popular circuit.

Counters are constructed such that the outputs of the lower-order flip-flops are used to clock the higher-order flip-flops.

The flip-flops do not all change state at the same time. They change state after the trailing edge of the input clock. These counters are asynchronous and are called *ripple counters*. A *synchronous counter* is a counter where all the flip-flops change state at the same time. Counters must be constructed with master-slave or edge-triggered flip-flops. In this chapter, all of the counters are made with JK master-slave flip-flops.

RIPPLE COUNTERS

A binary UP counter and its timing diagram are shown in *Figure 11-1*. The J and K inputs of all the flip-flops are high; therefore, each flip-flop is a toggle flip-flop. Each flip-flop changes state on the trailing edge of a clock pulse, as shown in the timing diagram. The count sequence of the UP counter is shown in *Figure 11-1*.

A binary DOWN counter and its timing diagram are shown in *Figure 11-2*. The $\overline{Q}$ output of each flip-flop provides the clock pulse of the next flip-flop. The flip-flops change state on the trailing edge of the clock pulse. The count sequence of the DOWN counter is also shown in *Figure 11-2*.

If a ripple counter has many stages, the propagation delays of the flip-flops may exceed the clock pulse period. The higher-order flip-flops may not change state correctly. The ripple counter is popular nonetheless because integrated-circuit flip-flops are fast enough for the higher-order flip-flops to change state properly. Ripple counters are used under the following conditions

(a) Where only the output of the last flip-flop is used, such as in a frequency divider.
(b) Where the flip-flop decoder combination drives a visual indicator or a control device whose slow response permits the device to ignore the ripple transients.
(c) Where the flip-flop outputs are strobed or sampled only after all rippling is completed.

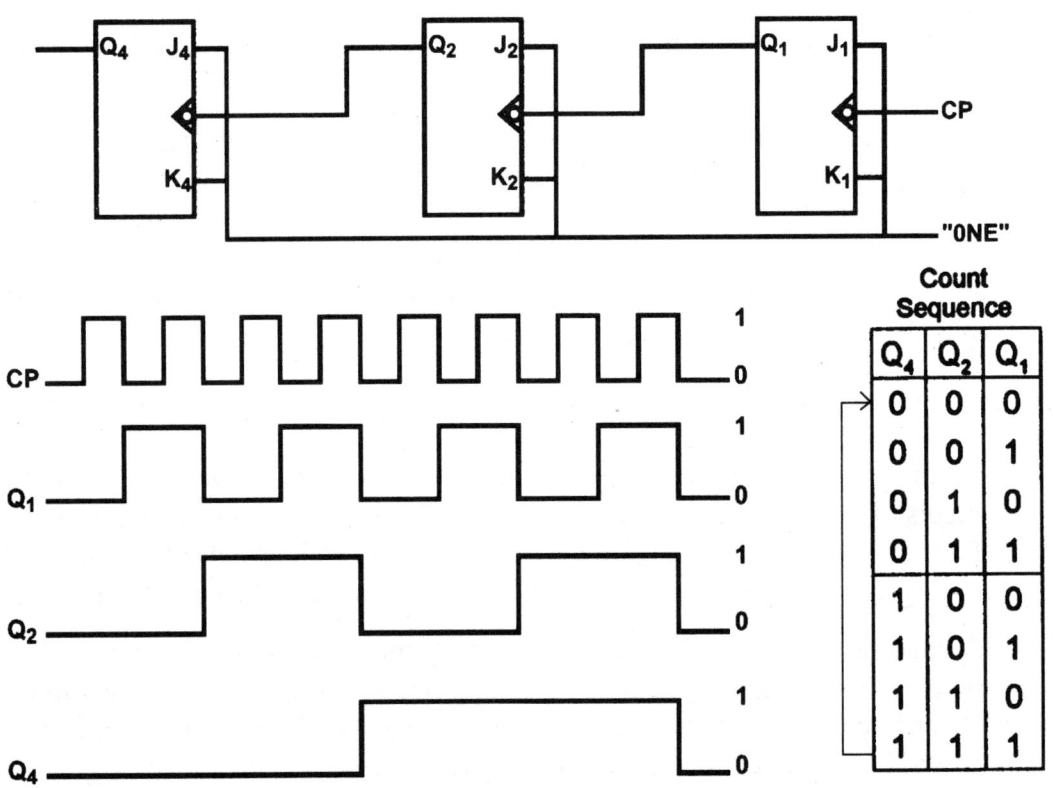

Figure 11-1. Binary-up counter.

SYNCHRONOUS COUNTERS

All of the flip-flops of a synchronous counter are clocked at the same time with the same clock pulse. The flip-flops therefore change state simultaneously with the same propagation delay.

A counter may have its change-of-state controlled through its J and K inputs. The flip-flops therefore do not change state at the same time; the counter has *pseudosynchronous operation*. A pseudosynchronous counter and its timing diagram are shown in *Figure 11-3*. The minimized present state of the counter is gated with the input clock to selectively trigger each flip-flop. Therefore, CP1 = CP, CP2 = Q1.CP, CP4 = Q2.Q1.CP and CP8 = Q4.Q2.Q1.CP

The counter is synchronous only if the gate delays are identical. This circuit is an improvement over the ripple counter because the difference between gate delays is less than even one flip-flop delay.

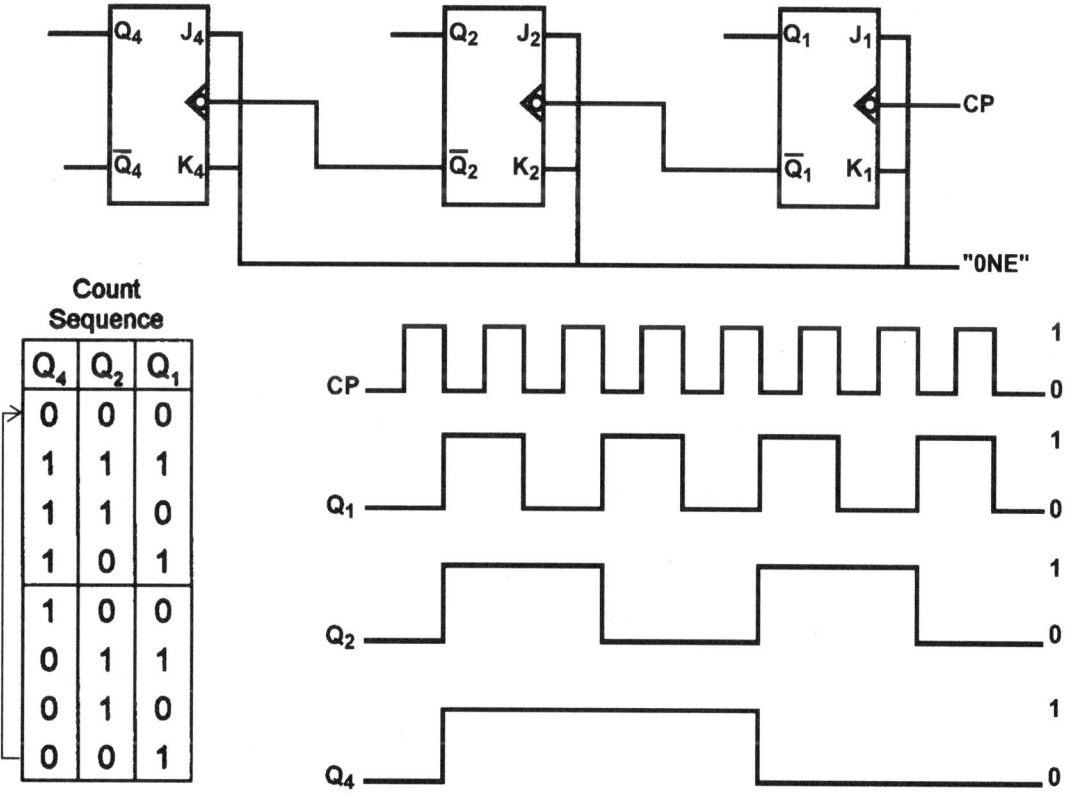

Figure 11-2. Binary-down counter.

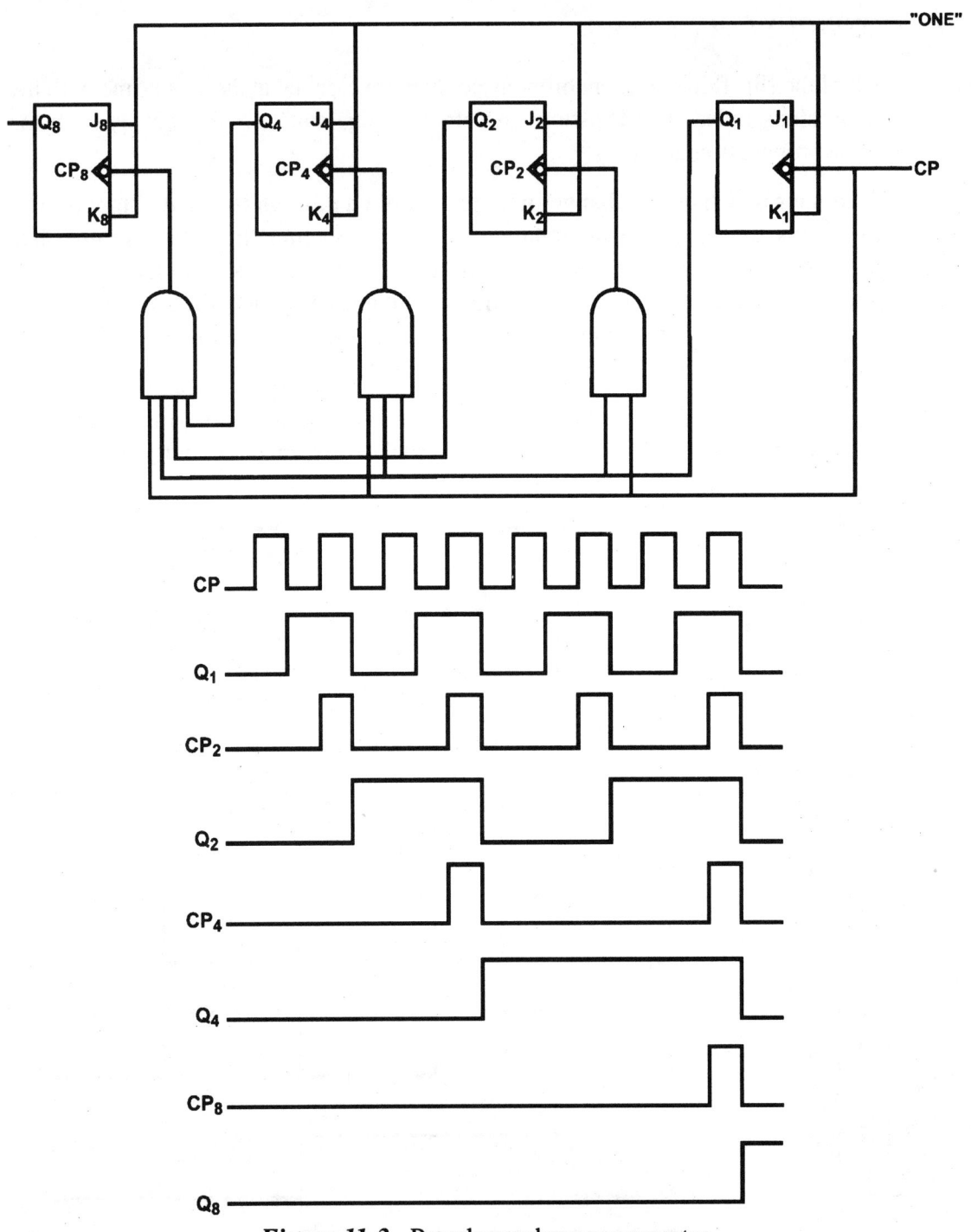

Figure 11-3. Pseudosynchronous counter.

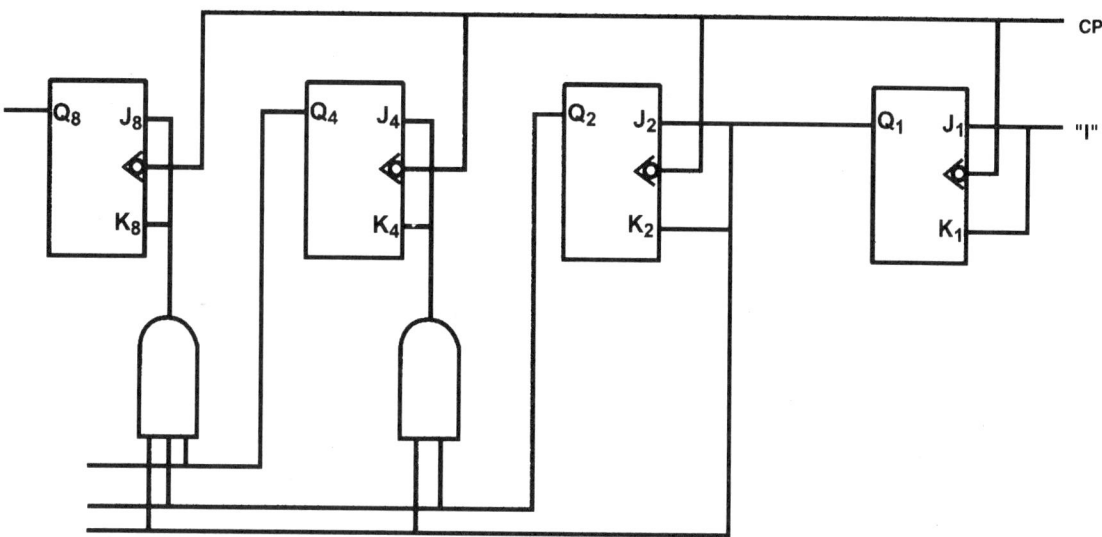

Figure 11-4. Synchronous counter with parallel carry.

The true synchronous counter has the same clock pulse applied to each flip-flop, and toggling is conditioned through the J and K inputs. A true synchronous counter is shown in *Figure 11-4*. It has parallel carry. In this circuit, $J1 = K1 = 1$, $J2 = K2 = Q1$, $J4 = K4 = Q2.Q1$, and $J8 = K8 = Q4.Q2.Q1$

A similar circuit with serial carry is shown in *Figure 11-5*. It reduces the fan-in requirements of the gates controlling the higher-order flip-flops. In this circuit, $J1 = K1 = 1$, $J2 = K2 = Q1$, $J4 = K4 = Q2.Q1 = Q2.J2$, and $J8 = K8 = Q4.Q2.Q1 = Q4.J4$

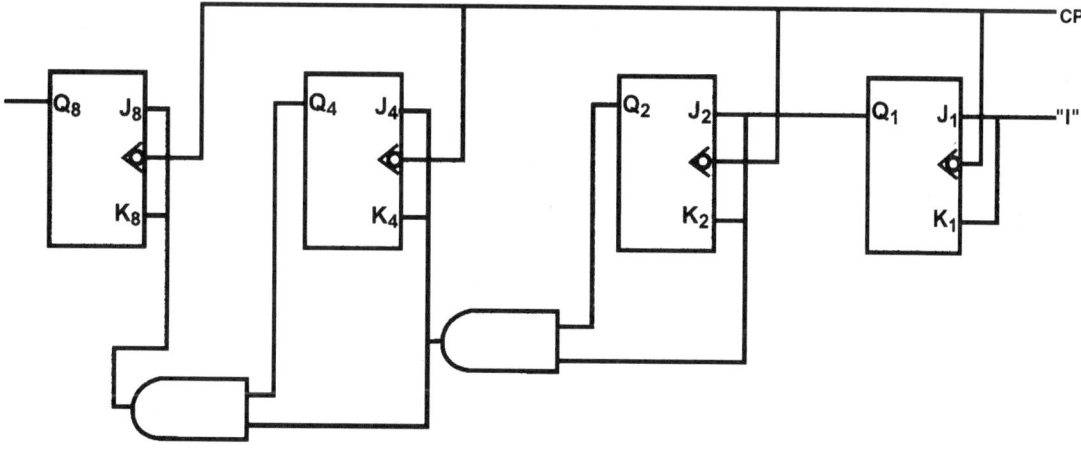

Figure 11-5. Synchronous counter with serial carry.

The maximum clock rate for a serial-carry counter is less than that for a parallel-carry counter because the JK input gates are cascaded. A good compromise is a serial carry connection between groups of parallel-carry connected flip-flops.

COUNTER DESIGN

A counter is designed by using the outputs available in the present state to enable the flip-flop changes required to produce the next state. Any master-slave or edge-triggered flip-flop may be used. *Table 11-1* lists the signals required at the inputs of different flip-flops to change from one state to the next state.

Present State	Next State	S	C	J	K	T	D
0	0	0	∅	0	∅	0	0
0	1	1	0	1	∅	1	1
1	0	0	1	∅	1	1	0
1	1	∅	0	∅	0	0	1

∅ = Don't care.

Table 11-1. Input signals for different flip-flops.

Example 11-1

It is required to design a counter using JK flip-flops with the following count sequence: 0, 7, 1, 3, 4, 2, 6, 0.

Table 11-2 shows the truth table for the required counter. To go from 0 (000) to 7 (111), each bit changes from a 0 to a 1. From *Table 11-1*, each J = 1 and each K is a "don't care" state; that is, it can be a 1 or a 0. This step is repeated for each state transition until *Table 11-2* is completed. A Karnaugh map is drawn and mapped for each J and K input function. The final counter circuit is shown in *Figure 11-6*. It should be noted that 5 is not in the required count sequence. It is treated as a "don't care" count state, simplifying the required counter design.

Example 11-2

To find the count sequence of the counter of *Example 11-1*, a truth table is prepared using the circuit of *Figure 11-6*. The present state column lists each possible count. The state of each J and K input is determined from the present state. The result is listed in *Table 11-3*. The count sequence is also shown in *Table 11-3*. Note that the "don't care" state 5 leads back to the count sequence at state 3. The next state is obtained from the J and K inputs of the present state.

Present State			Next State			J_4	K_4	J_2	K_2	J_1	K_1
Q_4	Q_2	Q_1	Q_4	Q_2	Q_1						
0	0	0	1	1	1	1	∅	1	∅	1	∅
1	1	1	0	0	1	∅	1	∅	1	∅	0
0	0	1	0	1	1	0	∅	1	∅	∅	0
0	1	1	1	0	0	1	∅	∅	1	∅	1
1	0	0	0	1	0	∅	1	1	∅	0	∅
0	1	0	1	1	0	1	∅	∅	0	0	∅
1	1	0	0	0	0	∅	1	∅	1	0	∅
1	0	1	∅	∅	∅	∅	∅	∅	∅	∅	∅

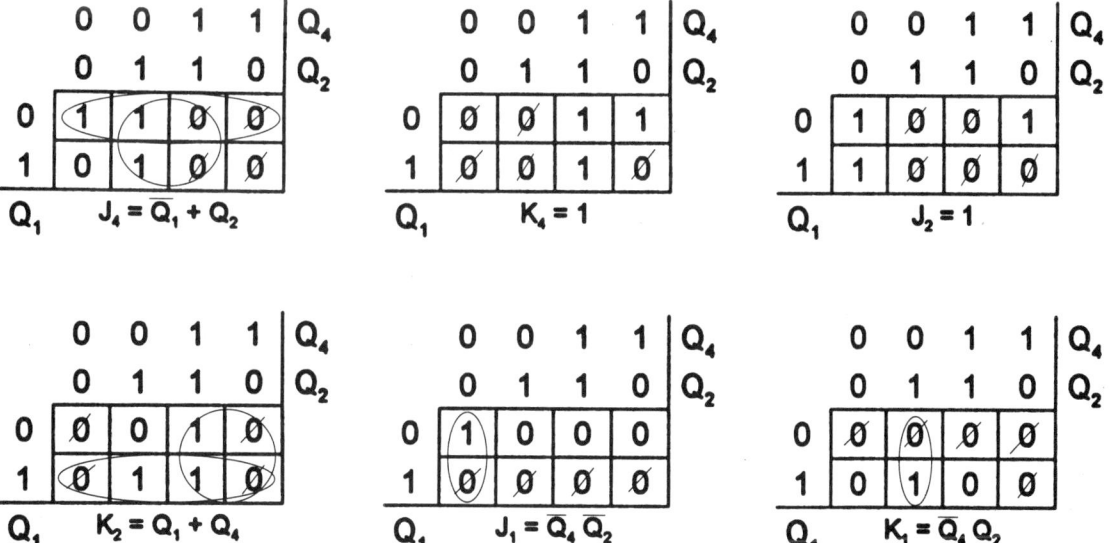

Table 11-2. Truth table for counter for *Example 11-1*.

82 / Digital Electronics

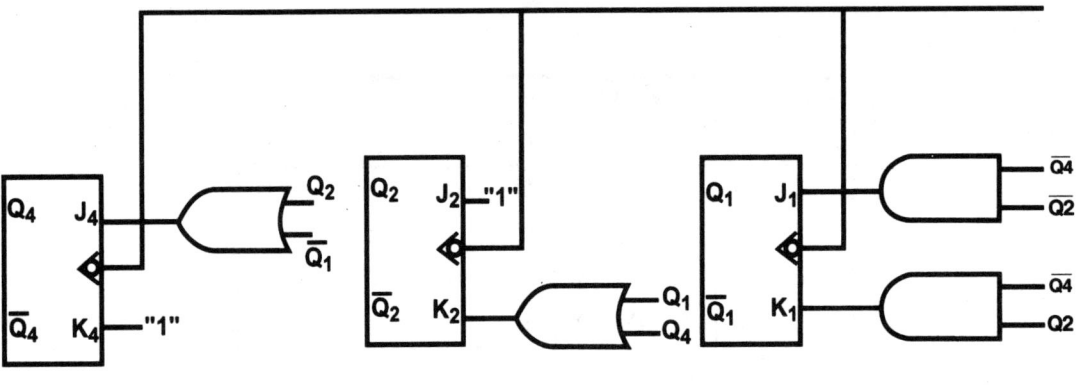

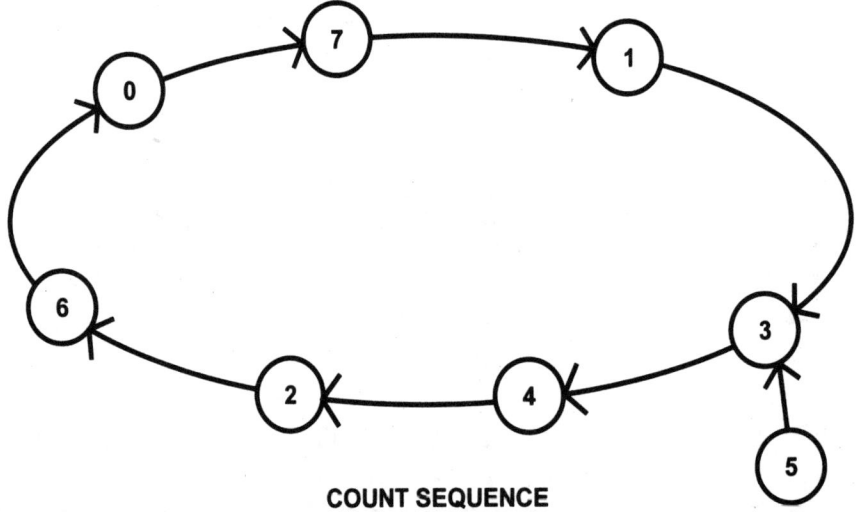

COUNT SEQUENCE

Figure 11-6. Counter circuit for *Example 11-1*.

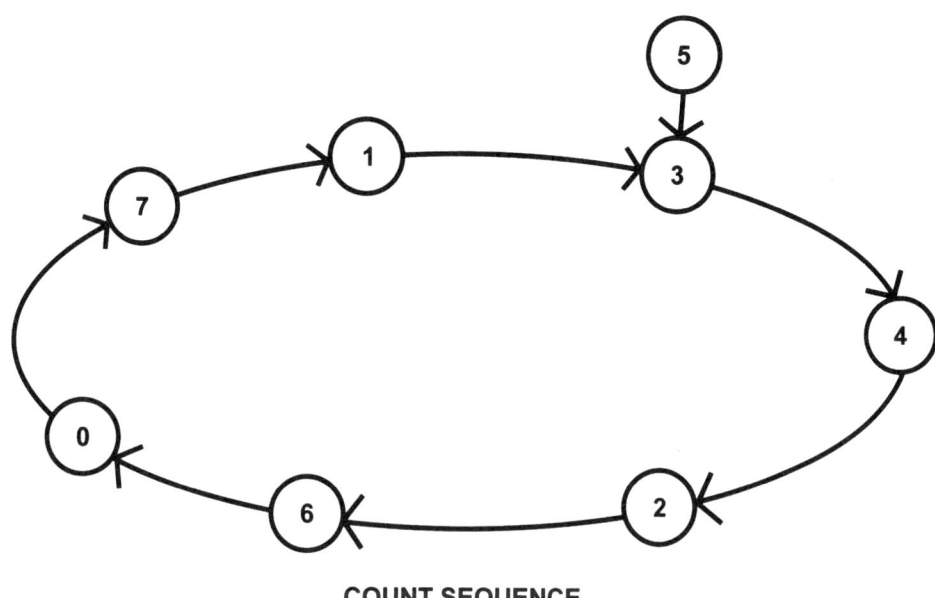

COUNT SEQUENCE

Present Status									Next Status		
Q_4	Q_2	Q_1	$J_4 = Q_2 + \overline{Q_1}$	$K_4 = 1$	$J_2 = 1$	$K_2 = Q_4 + Q_1$	$J_1 = \overline{Q_4}\overline{Q_2}$	$K_1 = Q_2\overline{Q_4}$	Q_4	Q_2	Q_1
0	0	0	1	1	1	0	1	0	1	1	1
0	0	1	0	1	1	1	1	0	0	1	1
0	1	0	1	1	1	0	0	1	1	1	0
0	1	1	1	1	1	1	0	1	1	0	0
1	0	0	1	1	1	1	0	0	0	1	0
1	0	1	0	1	1	1	0	0	0	1	1
1	1	0	1	1	1	1	0	0	0	0	0
1	1	1	1	1	1	1	0	0	0	0	1

Table 11-3. Truth table for *Example 11-2*.

PROBLEMS

Problem 11-1.
How many flip-flops are required for a binary counter that repeats its count sequence every eight input pulses?

Problem 11-2.
How does a ripple counter operate?

Problem 11-3.
What is the disadvantage of a ripple counter?

Problem 11-4.
For what applications are ripple counters used? Why?

Problem 11-5.
How does a synchronous counter operate?

Problem 11-6.
What is pseudosynchronous operation?

Problem 11-7.
When is a pseudosynchronous counter a synchronous counter?

Problem 11-8.
Discuss serial carry counters.

Problem 11-9.
Discuss parallel carry counters.

Problem 11-10.
Design a counter with the following count sequence using JK flip-flops: 1, 3, 4, 2, 6, 0, 7, 1.

Problem 11-11.
Verify the count sequence of the counter designed in ***Problem 11-10***.

Chapter Twelve
ARITHMETIC CIRCUITS

The half adder adds two bits (X and Y) to get a sum (S) and a carry (Co), according to the truth table given in *Table 12-1*. A block diagram and a logic circuit for the half adder are shown in *Figure 12-1*. The half adder does not consider any previous sums.

The full adder takes into account previous carries (Ci). The truth table for a full adder is given in *Table 12-2*. A block diagram and a logic circuit for the full adder are shown in *Figure 12-2*. The full adder may be designed using half adders, as shown in *Figure 12-3*. A carry is generated in a full adder stage if both X and Y are 1. The carry is propagated through the stage if either (but not both) X or Y is 1 and Ci is also 1.

RIPPLE ADDER

A ripple adder consists of a number of stages of full adders, such that the carry out of the *ith* stage is the carry into the *(i+1)th* stage. The time required to perform addition in the ripple adder is the time required for the propagation of the carries in the stages.

x	y	S	C_o
0	0	0	0
0	1	1	0
1	0	1	0
1	1	0	1

$S = x \oplus y$
$C_o = x \cdot y$

Table 12-1. Half-adder truth table.

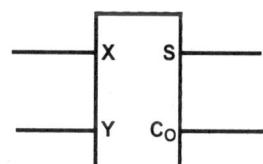

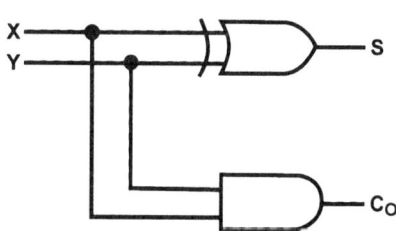

Figure 12-1. Half adder block diagram and logic circuit.

86 / Digital Electronics

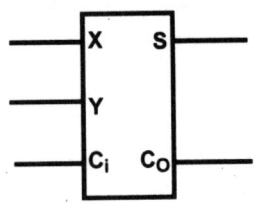

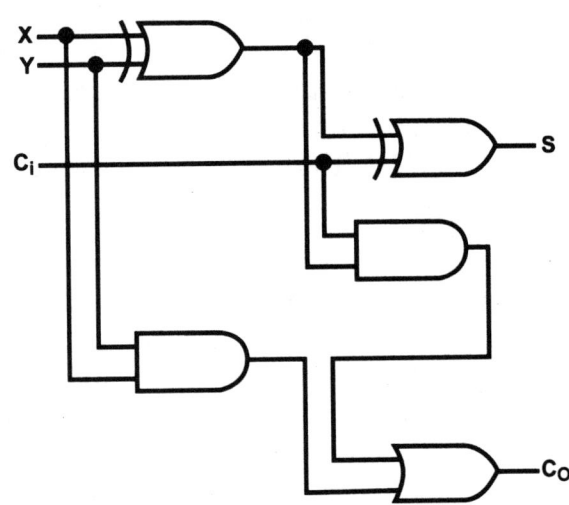

x	y	C_i	S	C_o
0	0	0	0	0
0	0	1	1	0
0	1	0	1	0
0	1	1	0	1
1	0	0	1	0
1	0	1	0	1
1	1	0	0	1
1	1	1	1	1

$$S = x \oplus y \oplus C_i$$
$$C_o = x \cdot y + (x \oplus y)C_i$$
Where $x \cdot y$ = Generated Carry
Where $x \oplus y$ = Propogated Function

Table 12-2. Full-adder truth table.

Figure 12-2. Full adder block diagram and logic circuit.

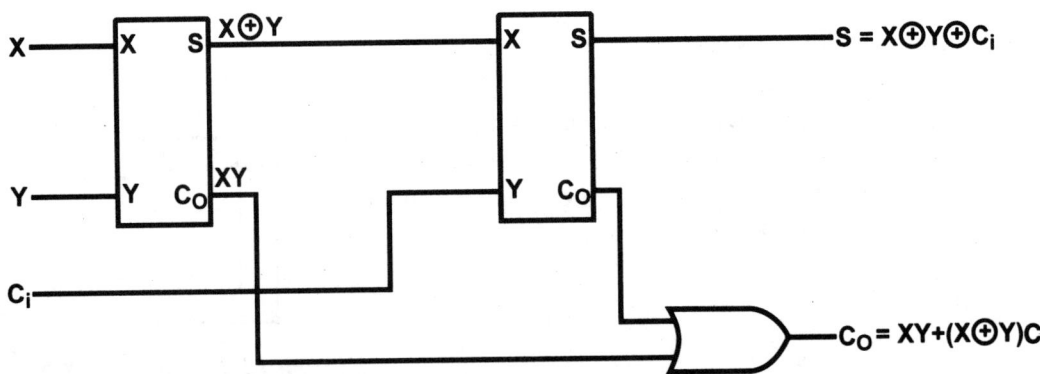

Figure 12-3. Full adder using half adders.

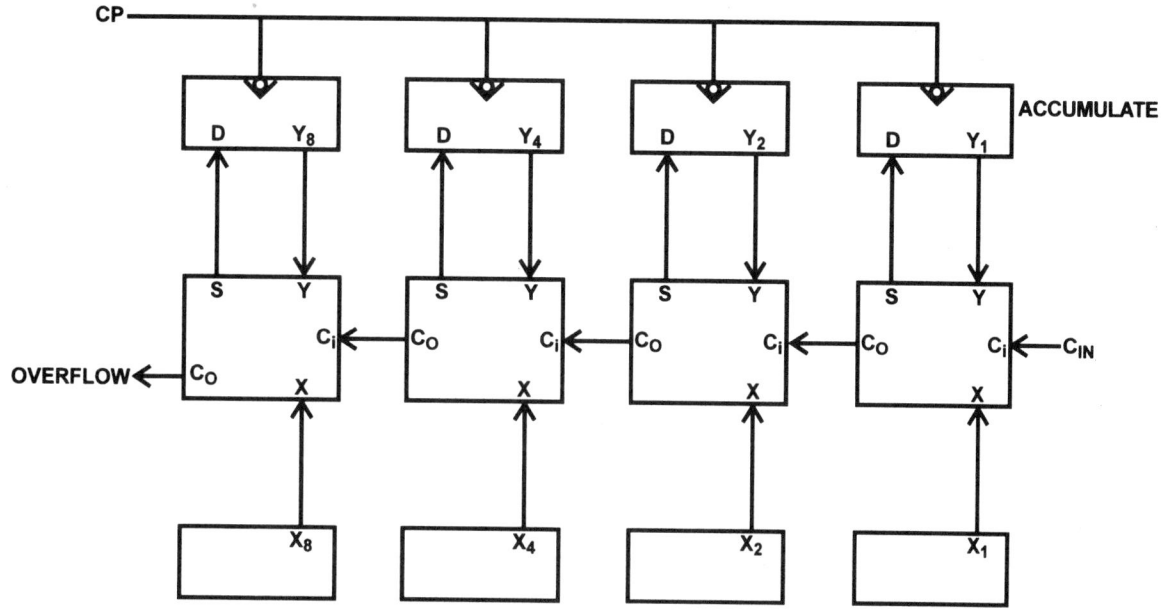

Figure 12-4. Block diagram of a ripple adder.

The carry will not propagate through all stages in every addition. However, the addition operation must have sufficient time allotted; that is, it must equal the longest propagation time plus the addition time of the last adder.

A block diagram for a ripple or parallel adder is shown in *Figure 12-4*. The sum is clocked into the accumulator on the trailing edge of the clock pulse.

The worst case addition time occurs when 0001 and 1111 are added together. The carry generated in the least significant stage must propagate (ripple) through each stage to the most significant stage. Sufficient time must be allowed for the worst case addition time before the sum can be clocked into the accumulator.

The accumulator must be constructed with synchronous flip-flops or registers. The X register may be made with either synchronous or asynchronous registers.

LOOK AHEAD ADDER

The look ahead, or anticipated carry adder does not wait for a carry to ripple through it. Instead, each stage of a look ahead adder examines the inputs of all previous stages to determine what its input carry will be. A block diagram and logic circuit of a four-bit block-anticipated carry adder are shown in *Figure 12-5*.

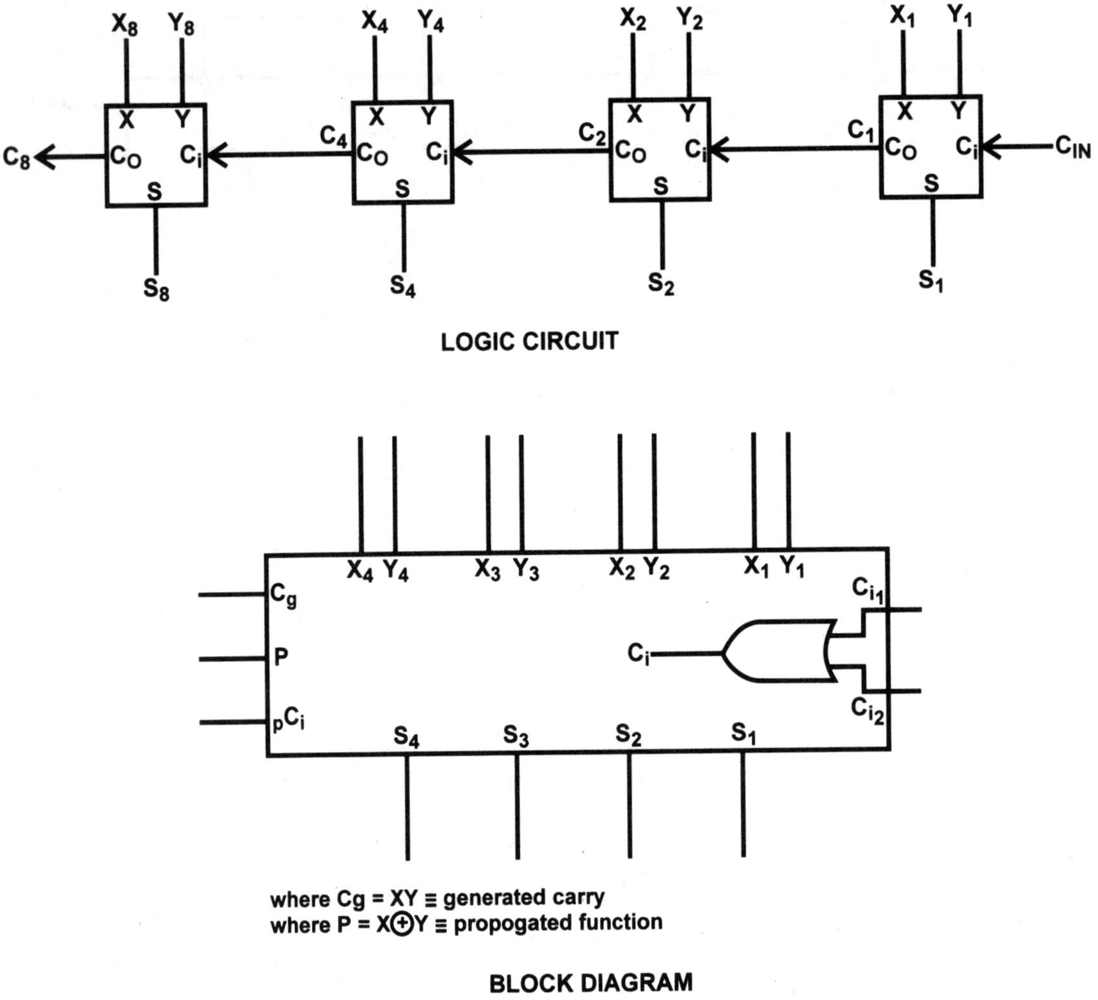

88 / Digital Electronics

LOGIC CIRCUIT

where $C_g = XY \equiv$ generated carry
where $P = X \oplus Y \equiv$ propogated function

BLOCK DIAGRAM

Figure 12-5. Block diagram and logic circuit of a four-bit look-ahead adder.

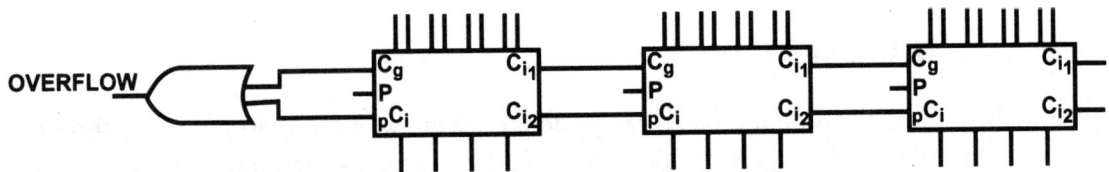

Figure 12-6. Character-to-character ripple adder.

APPLICATIONS

The character-to-character ripple adder is the slowest application of a ripple adder. The look-ahead principle is applied within each block. The carry input to each four-stage section examines only the carry output of the preceding section, so that each Cg output must ripple through the higher order sections. A character-to-character ripple adder is shown in *Figure 12-6*.

In a fast adder system, the carry generated in or propagated through the least significant section is passed directly to the second carry input of the most significant section, if the propagation functions of the other stages are 1. A fast adder system is shown in *Figure 12-7*. A carry generated in section 2 is fed to the most significant section if the propagation function of section 3 is 1.

A converter from sign and magnitude to signed-twos complement is shown in *Figure 12-8*.

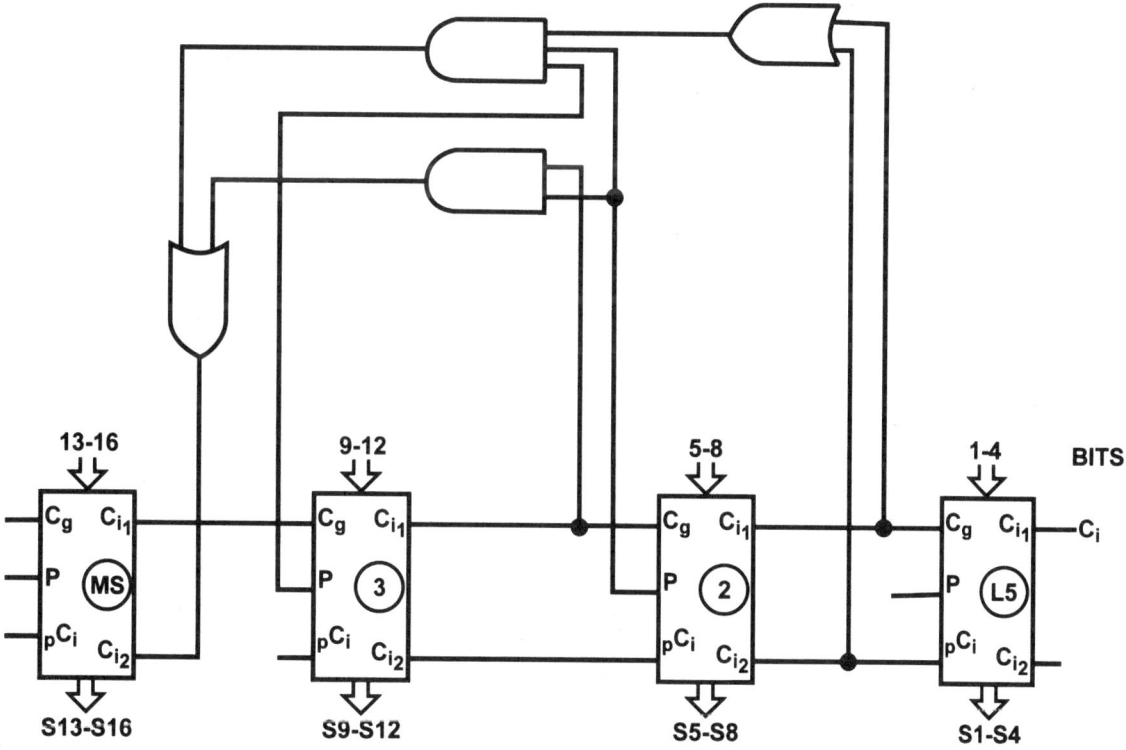

Figure 12-7. Faster adder system.

90 / *Digital Electronics*

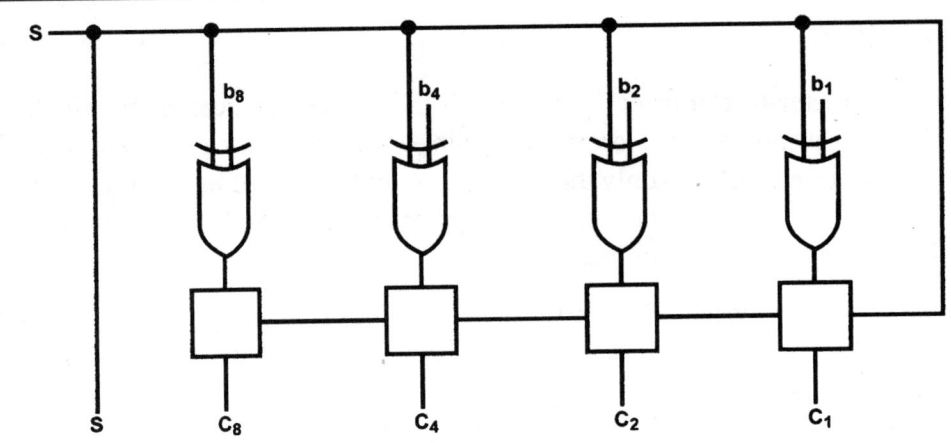

Figure 12-8. Sign and magnitude to signed-twos complement converter.

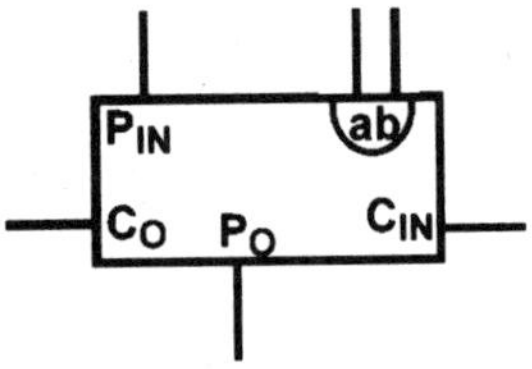

BLOCK DIAGRAM

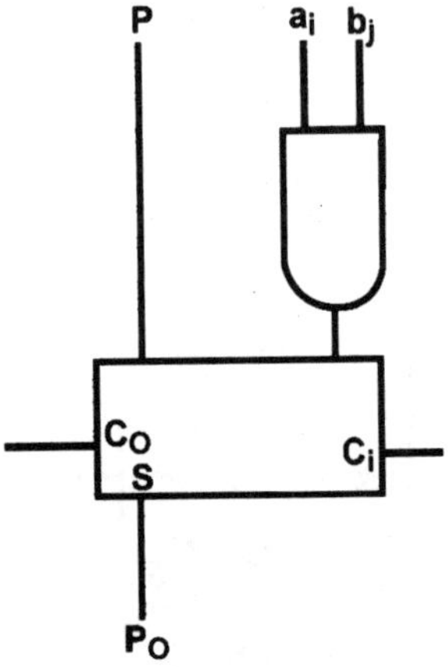

LOGIC CIRCUIT

Figure 12-9. Basic bit multiplier block diagram and logic circuit.

A	B	P
0	0	0
0	1	0
1	0	0
1	1	1

$P = A \cdot B$

				A_4	A_2	A_1
			×	B_4	B_2	B_1
				A_4B_1	A_2B_1	A_1B_1
				P_4	P_2	P_1
			A_4B_2	A_2B_2	A_1B_2	
		P_{16}	P'_8	P'_4	P'_2	P_1
		A_4B_4	A_2B_4	A_1B_4		
	P_{32}	P_{16}	P_8	P_4	P_2	P_1

Table 12-3. Multiplier truth table.

The truth table for a multiplier is given in *Table 12-3*. The multiplication steps are also shown in *Table 12-3*. A basic-bit multiplier with an accumulation of a partial product block diagram and a logic circuit are shown in *Figure 12-9*. A three-bit multiplier may be designed using several basic bit multipliers with the accumulation of partial product, as shown in *Figure 12-10*.

PROBLEMS

Problem 12-1.

What is the main difference between the half adder and the full adder?

Problem 12-2.

What is the worst case addition time of a ripple adder?

Problem 12-3.

A ripple adder accumulator must be made with what type of register(s)?

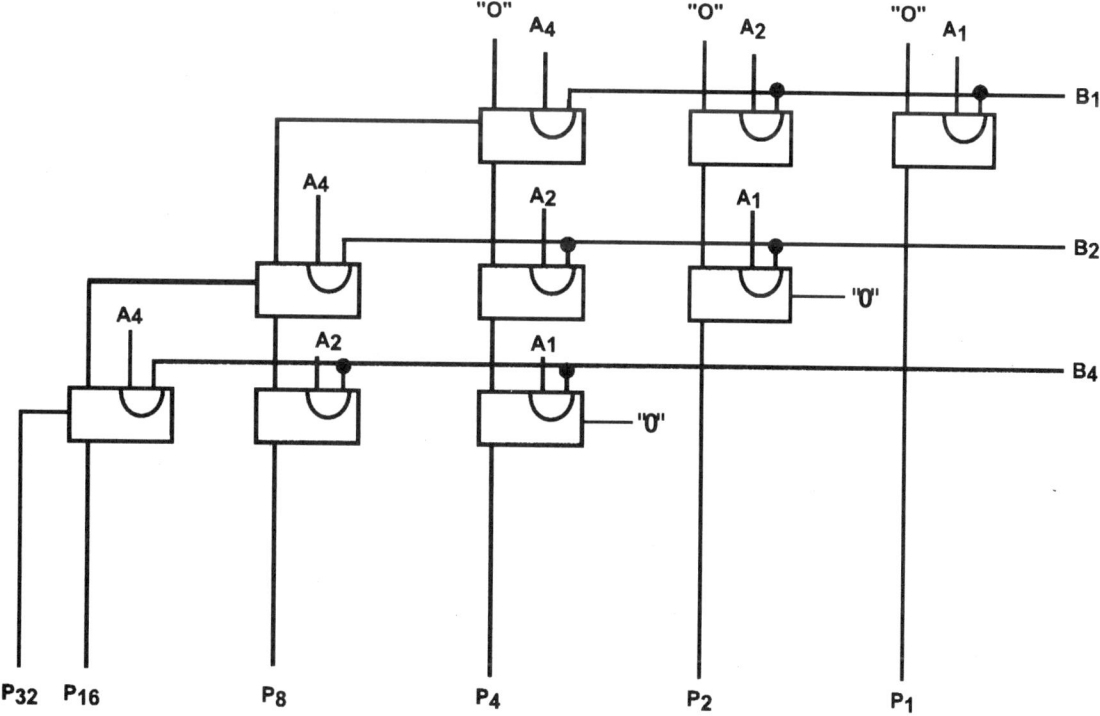

Figure 12-10. Three-bit multiplier.

Problem 12-4.

What is the main difference between a ripple adder and a look ahead adder?

Problem 12-5.

How does a look ahead adder function?

Problem 12-6.

What is the slowest application of a ripple adder?

Problem 12-7.

How does a fast adder function?

Problem 12-8.

A sign and magnitude to signed-twos complement converter is shown in *Figure 12-8*. If S = 1 and b8b4b2b1 = 0100, what is c8c4c2c1?

Problem 12-9.

Multiply 011 by 101.

Chapter Thirteen
MEMORY

Memory is made from two-state devices and is used for the storage of digital data. There are internal memories and external memories.

Internal memory is an integral part of the system. There is no need for system input/output procedures for the transfer of data. Fast local storage is provided by register memories. Slower mass memory may be used for the storage of programs and data. The ROM and RAM memories of a computer are internal memories.

External memory is not an integral part of the system, and therefore requires input/output procedures for the transfer of data. External memory is slower than internal memory. A diskette is an external memory device.

Writing into memory is placing temporary information into memory.

Programming is the permanent placement of information into a blank (read-only) memory.

Reading is the interrogation of a selected portion of the memory. Reading the whole memory is known as *dumping*.

Volatile memory loses data when power is removed from the memory. The internal memory of a computer is volatile.

Non-volatile memory does not loose data when power is removed from the memory. A diskette is a non-volatile memory.

Address is the location in memory of a word, byte, or bit; n address inputs provide 2^n addresses.

Access is the scheme for making addresses available.

Access time is the time it takes for data to be available once a request for data has been recognized.

Memory cycle time is the minimum interval between the reading of two words.

Random access memory is a memory in which the address selection may be made in any order. It is also known as RAM.

Dynamic memory, unlike *static memory*, cannot retain data indefinitely because the information is stored as electrical charge in small capacitors. The leakage resistance across the capacitor provides a discharge path for the stored charge. The memory capacitors must be periodically recharged in order to retain the data. A memory refresh occurs every one or two milliseconds.

Sequential address may be linear such as on a magnetic tape, or cyclic such as on a long shift register.

Read-only memory is a memory used only for the permanent storage of data. It usually has no rewriting capability. It is also known as ROM. A ROM memory that is programmable by the manufacturer is a *programmable read-only memory* (PROM). An EAROM is a ROM that can be erased and reprogrammed by a special process.

A *memory address register* holds the address currently selected.

A *memory data register* passes all data written into and read from the memory.

There are two types of memory systems; *word-oriented memory* systems and *bit-oriented memory* systems.

The word-oriented system is two-dimensional, and it is also known as a *linear select system*. A linear select circuit is shown in *Figure 13-1*. There is a sense line (for reading) connected to the memory data register input and a write line connected to the memory data output. Semiconductor memories usually have a chip or memory enable input (ME) which enables the read/write circuitry of the integrated circuit. The word-oriented system is fast but the size and cost of the decoder is excessive for larger memories.

The bit-oriented memory system is three-dimensional and is also known as *coincident signal select*. A bit-oriented circuit is shown in *Figure 13-2*. Two signals are required to select any bit, one from each of two decoders. The final "ANDing" in the address decoding process is provided by the coincidence of the X and Y selection signals. One bit is stored at each intersection of the X and Y select lines. The memory plane can store XY one-bit words. An n bit address is partitioned such that $nX = nY = n/2$, which yields $2^{n/2}X$ select and $2^{n/2}Y$ select lines. This requires two (one out of $2^{n/2}$) decoders instead of the single (one out of 2^n) with a word-oriented linear select system. This is a tremendous saving for larger memory systems.

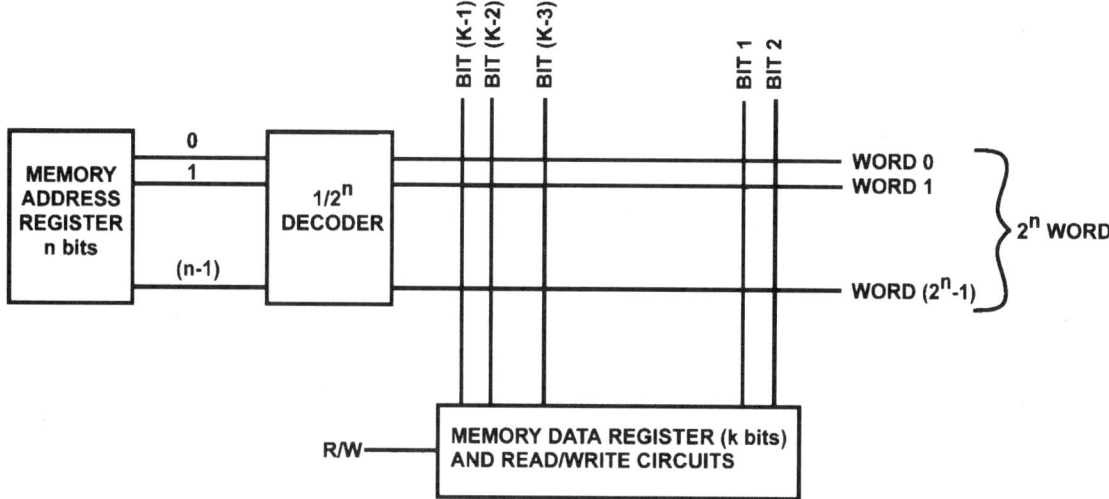

Figure 13-1. Word-oriented system.

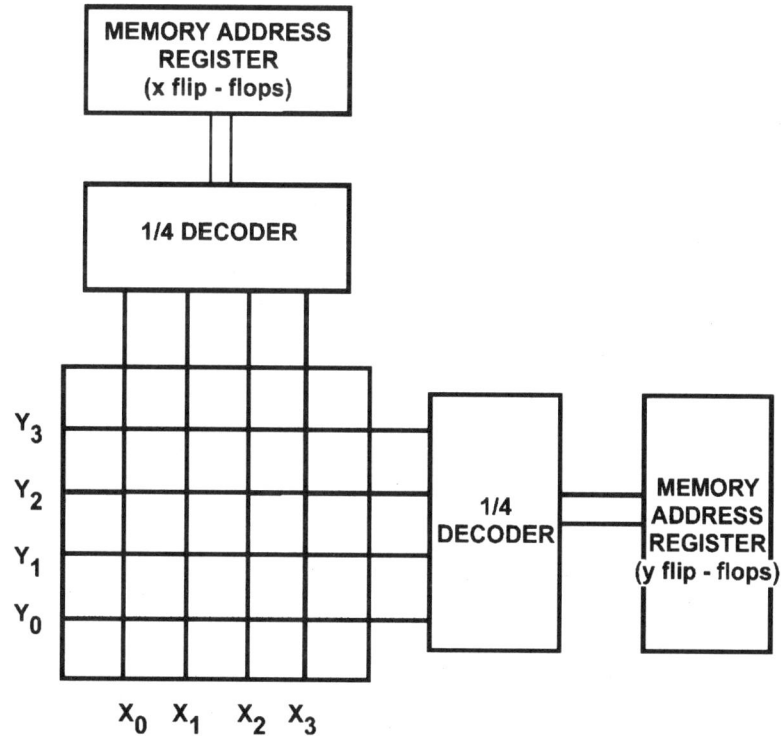

Figure 13-2. Bit-oriented system.

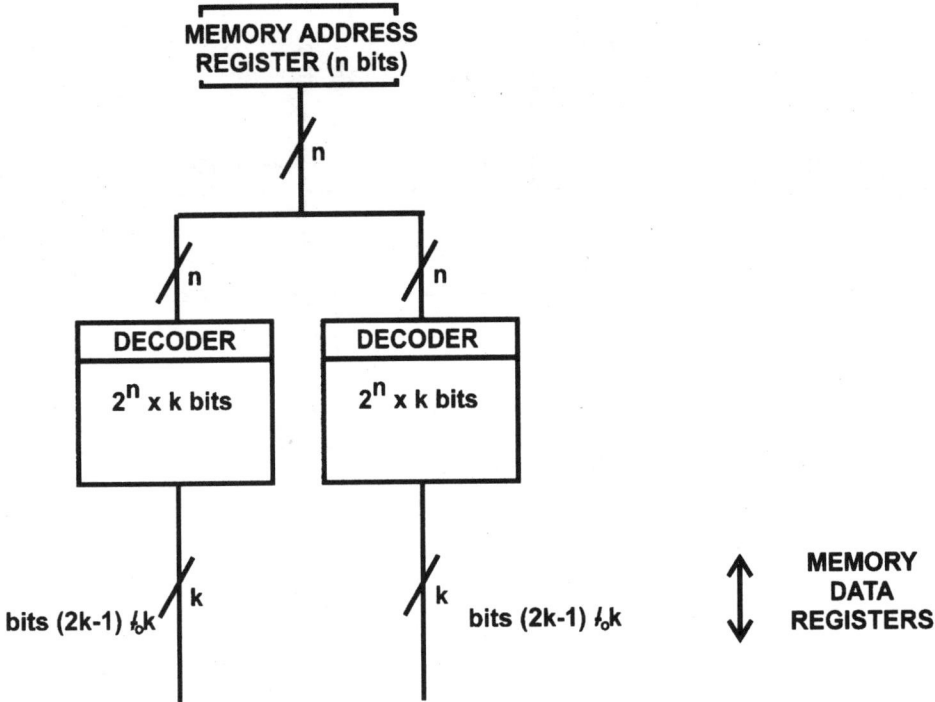

Figure 13-3. Circuit for expanding the word size of a memory.

If $n = 12$, a linear-select memory requires a 1/4096 decoder and 4096 select lines. A coincident selection memory requires $nX = nY = 6$, such that two 1/64 decoders are required to drive 128 select lines.

MEMORY EXPANSION

Memory can be expanded to increase word size, to increase the number of words, and to change the format of a ROM.

The word size of a memory can be increased by using the circuit shown in *Figure 13-3*. The memory stores 2^n words each with $2k$ bits. The number of words or bits remains the same.

The number of words of a memory can be increased by using the circuit shown in *Figure 13-4*. The memory stores $2^{n+2} \times k$ bit words; that is, in *Figure 13-4*, $4 \times 2^n \times k$ bit words.

The ROM format may be changed by using the circuit shown in *Figure 13-5*. A 32 x 8 bit ROM is being used as a 64 x 4 bit ROM.

Some common memory types and specifications are compared in *Table 13-1*.

PROBLEMS

Problem 13-1.

Compare internal and external memory.

Problem 13-2.

How many addresses can six address inputs provide?

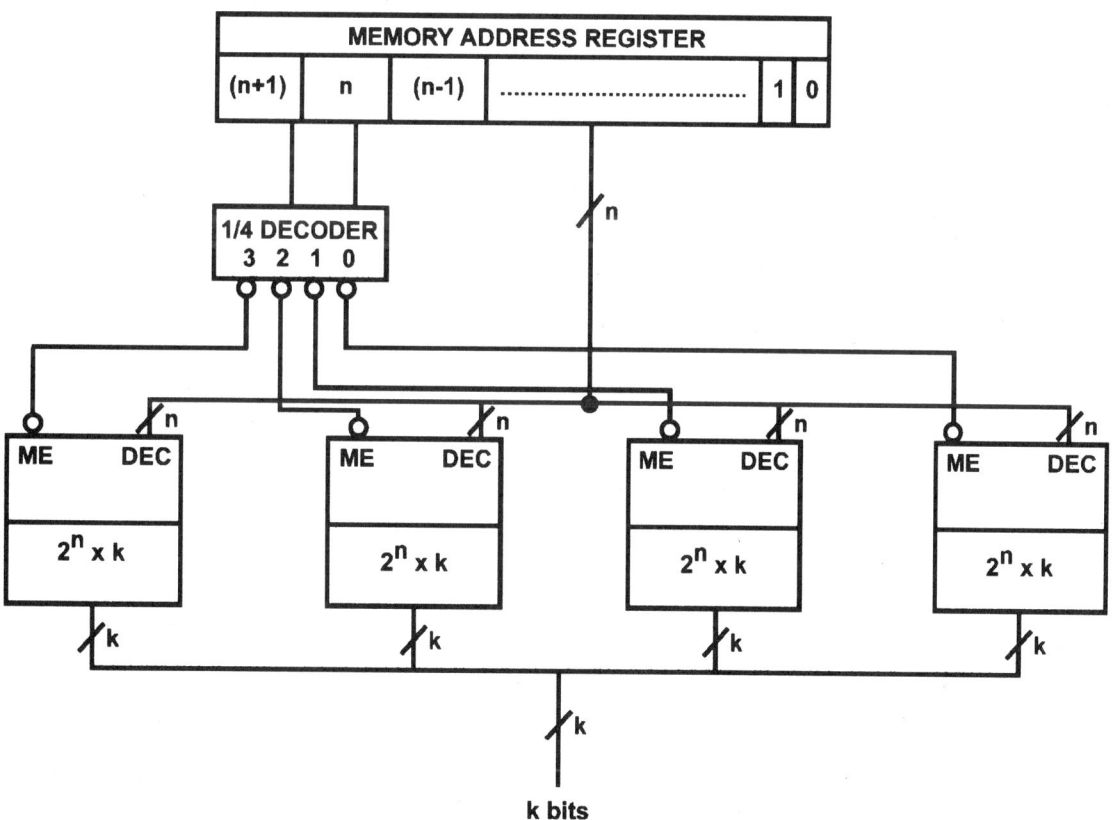

Figure 13-4. Circuit for increasing the number of words for a memory.

Problem 13-3.

What are access time and memory cycle time?

Problem 13-4.

Compare RAM and ROM.

Problem 13-5.

What is a PROM?

Problem 13-6.

What is an EAROM?

Problem 13-7.

Compare word-oriented and bit-oriented memory systems.

Problem 13-8.

In a bit-oriented system, how many X select lines and how many Y select lines are there if the system has a 5 bit address?

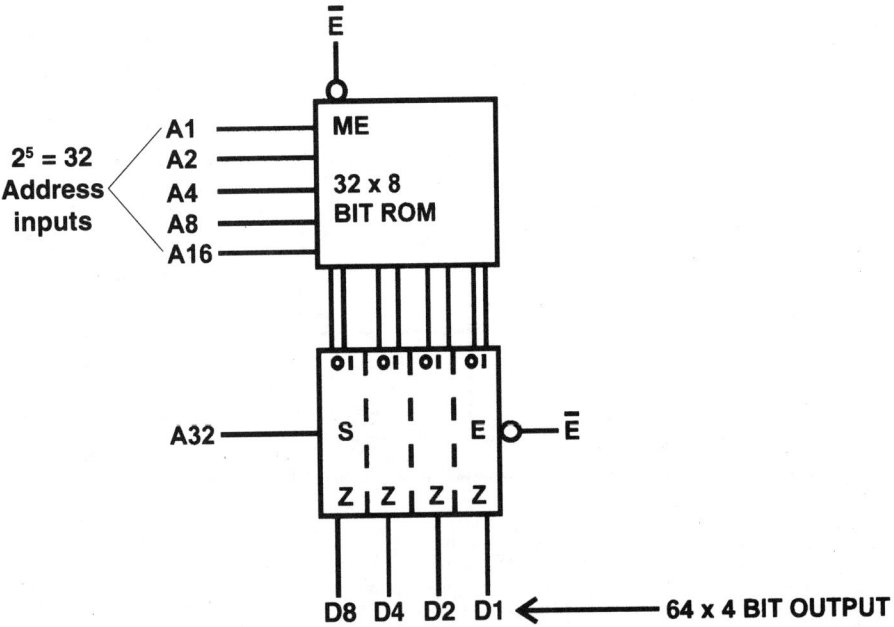

Figure 13-5. Circuit for changing the ROM format.

Problem 13-9.

How many decoders, and what type of decoders, are required for a bit-oriented system if it has a 5-bit address?

Problem 13-10.

How many decoders, and what type of decoders, are required for a word-oriented system if it has a 5-bit address?

Problem 13-11.

Expand the word size of a memory by a factor of 4.

Problem 13-12.

Increase the number of words of a memory by a factor of 4.

Problem 13-13.

Change the format of a ROM from 64 x 2 bit to 128 x 1 bit.

Type	Typical Capacity	Access Mode	Access Time	Volatile	Destructive Readout	Organization	Use
Semiconductor (ms1, Ls1) Bipolar	10^6 bits/in^2	Random (RAM) and Read Only (ROM)	100 - 500 ns.	Yes	No	Coincident	Local, Internal (Scratch Pad)
FET						Word-Oriented	
Core	10K - 100K	Random	1 - 10 ns.	No	Yes	1. Coincident 2. Word-Oriented	Main, Internal
Thin Film Spots Plated Wire	500K	Random	100 ns.	No	Yes	Word-Oriented	Main, Internal
Delay Line		Sequential (Cyclic)		Yes	No	"	Special
Drum	20K - 5m	Sequential (Cyclic)	10 - 40 ms.	No	No	"	Internal/External
Disk	1 - 20 x 10^6	Sequential (Cyclic)	10 - 70 ms.	No	No	"	Internal/External
Magnetic Tape	2 - 50 x 10^6 Per Reel	Sequential (Progressive)	1 - 300 sec.	No	No	"	Bulk External

Table 13-1. Common memory types and specifications.

Chapter Fourteen
DIGITAL-TO-ANALOG & ANALOG-TO-DIGITAL CONVERTERS

Signals in the real world are *analog*. An analog signal can have any value; it is not restricted to discrete values as are digital signals. Temperature, motor speed, and pressure are analog signals that may be operated upon by a digital system. The analog signal must be converted to a current or voltage by a transducer; the signal is then converted to a *digital signal* by an analog-to-digital (A/D) converter. The digital system operates on the digital signal. The new digital signal is converted to an analog signal by a digital-to-analog (D/A) converter. The final analog signal can be converted by another transducer to a different type of analog signal. *Figure 14-1* is a block diagram of a typical digital system.

DIGITAL-TO-ANALOG CONVERTERS

A D/A converter weighs each bit of the input digital signal according to its position, and sums the results. The D/A converter operates on the input current. Therefore, the basic output of a D/A converter is a *current*. The time required for the D/A converter output to arrive at and stay within one bit of its stable value is known as its *settling time*.

There are two types of D/A converter circuits, based on current output resistor ladders: the *binary weighted ladder* and the *R-2R ladder*.

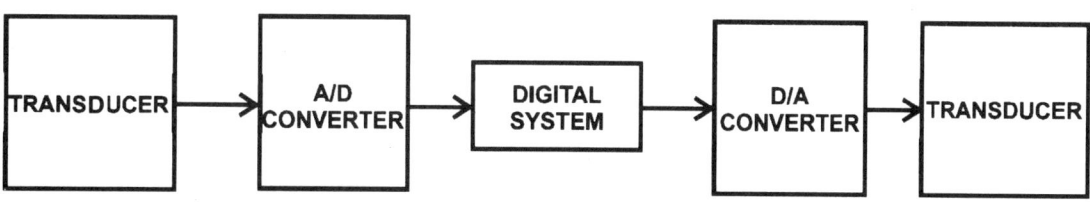

Figure 14-1. Block diagram of a digital system.

102 / *Digital Electronics*

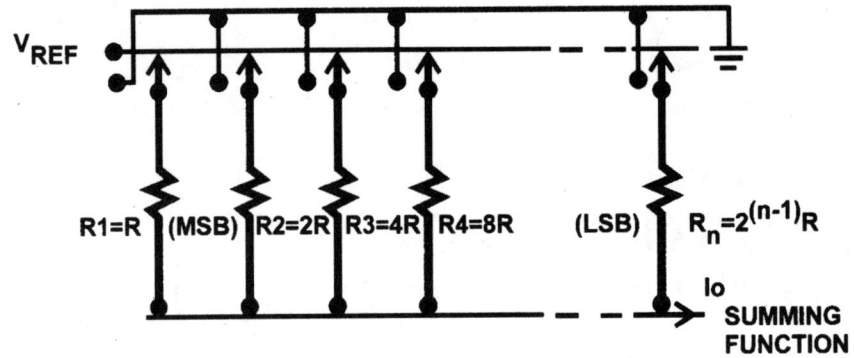

Figure 14-2. Binary-weighted ladder D/A converter.

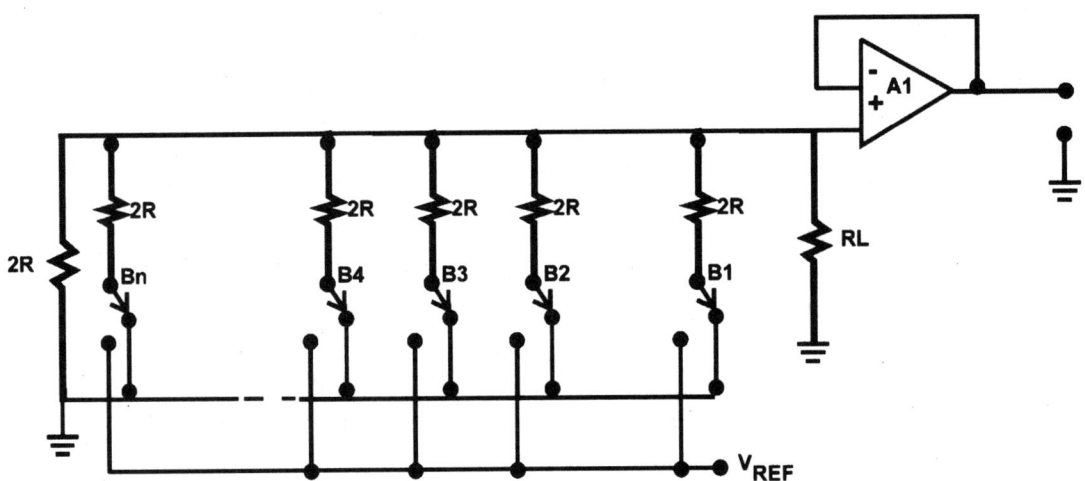

Figure 14-3. R-2R ladder D/A converter.

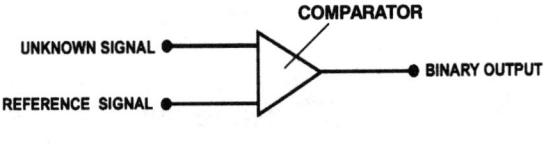

Figure 14-4. Block diagram of an A/D converter.

Binary Weighted Ladder D/A Converter

A binary weighted ladder D/A converter is shown in *Figure 14-2*. In this circuit, R1 = R, R2 = 2R, R3 = 4R, and Rn = $2^{(n-1)}$R.

The output current of the D/A converter is Io = Vref(a1/R + a2/2R +...+ an/$2^{(n-1)}$R), where *a1* is the MSB of the binary code and *an* is the LSB of the binary code.

If the binary weighted resistor ladder is more than 8 bits long, the resistor values at the ends of the ladder become either very large or very small. The R-2R ladder does not have this problem.

R-2R Ladder D/A Converter

All of the resistors in an R-2R ladder have a value of either R or 2R. If the load resistor is much larger than 2R, then the output voltage is Eo = Vref(a1/2 + a2/4 + a3/8 +...+ an/2^n), where *a1* is the MSB of the binary code and *an* is the LSB of the binary code. *Figure 14-3* is the circuit of an R-2R ladder D/A converter.

The full scale output voltage depends upon the reference voltage (Vref) and the bit length (n). It is Eo(fs) = Vref(2^n- 1)/2^n.

ANALOG-TO-DIGITAL CONVERTERS

The analog-to-digital converter is an analog comparator; it compares an unknown analog signal to a reference analog signal. If the unknown signal is less than the reference analog signal, the comparator output is LOW. The comparator output is HIGH when the unknown signal exceeds the reference signal. *Figure 14-4* shows a block diagram of an A/D converter.

Analog-to-digital converters can use several methods to compare the unknown analog signal and the reference analog signals: *parallel conversion*, *successive approximation conversion*, *counter conversion* and *dual slope conversion*.

Parallel A/D Converter

In a parallel A/D converter, the input signal is compared to several reference signals by as many comparators. The comparator outputs are decoded by a logic decoding network, as shown in *Figure 14-5*. The comparator output goes LOW when the input signal is less than or equal to the reference signal. Otherwise, it goes HIGH.

Parallel A/D converters are very fast but they require many comparators: for n bits they require (2^n-1) comparators.

Successive Approximation A/D Converter

The successive approximation converter is the most popular type of A/D converter. The input signal is compared with a D/A converter output, one bit at a time, as shown in *Figure 14-6*. A control circuit drives a D/A converter whose output is fed to a comparator. When the START signal occurs, the control circuit sends an MSB signal of *1* to the D/A converter, which sends a signal of one-half the full scale signal to the comparator. If the input exceeds 1/2 FS, the MSB is left on; otherwise, it is reset to 0. In both cases, the next bit is tried. The process is repeated with each bit. When the LSB comparison is completed, the control circuit sends a STATUS signal to the output register to indicate that the binary output is available.

Successive approximation converters have high resolution, high speed and accuracy. Serial output data may be obtained, if required, MSB first. The conversion time of a successive approximation converter is nT, where n is the number of bits in the binary word and T is one clock pulse period.

Dual Slope A/D Converter

The dual slope A/D converter converts voltage to time with an integrating network. The elapsed time is measured with a counter, as shown in *Figure 14-7*. Either the unknown signal or a negative reference signal is applied to the input of the integrator. The integrator output is applied to the comparator, while the other terminal of the comparator is grounded. When the unknown input signal is connected to the integrator, conversion begins. The comparator turns on when the integrator output exceeds a small threshold voltage. When the output of the integrator becomes positive enough to trigger the comparator, the comparator output starts the counter counting a predetermined number of clock pulses (T1). When this number is reached, the comparator input is connected to the negative reference voltage and the counter is reset to 0. The integrator integrates the negative reference voltage until it exceeds the comparator threshold. The counter is stopped at a holding number (T2). The unknown input voltage is Vin = (T2/T1) x reference voltage.

Digital-to-Analog & Analog-to-Digital Converters / **105**

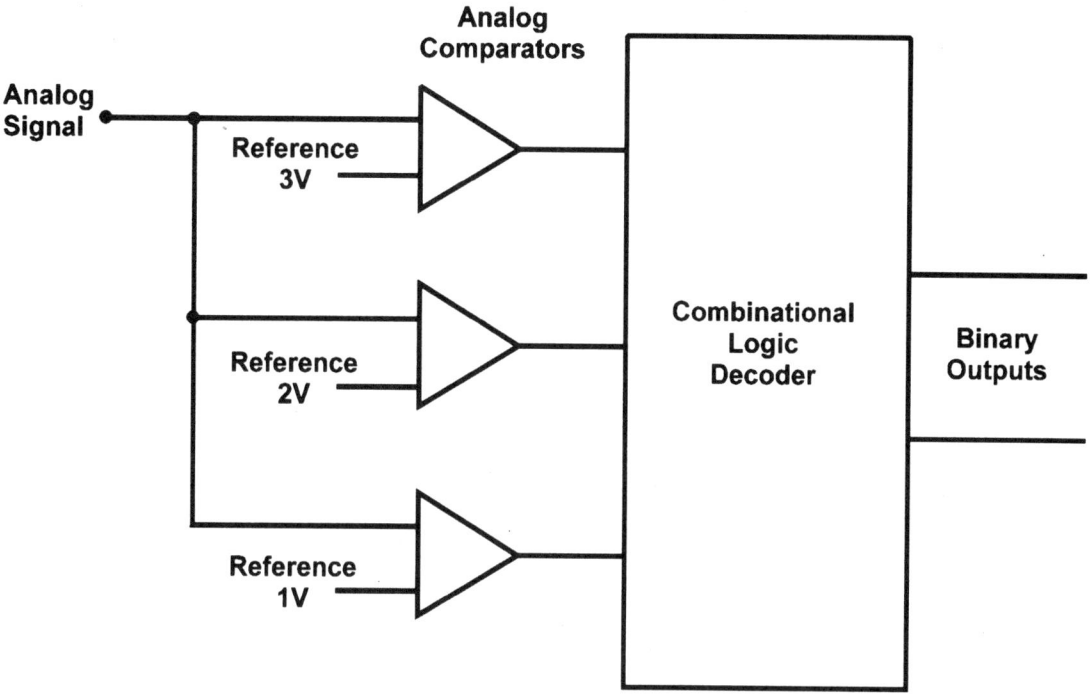

Figure 14-5. Parallel A/D converter.

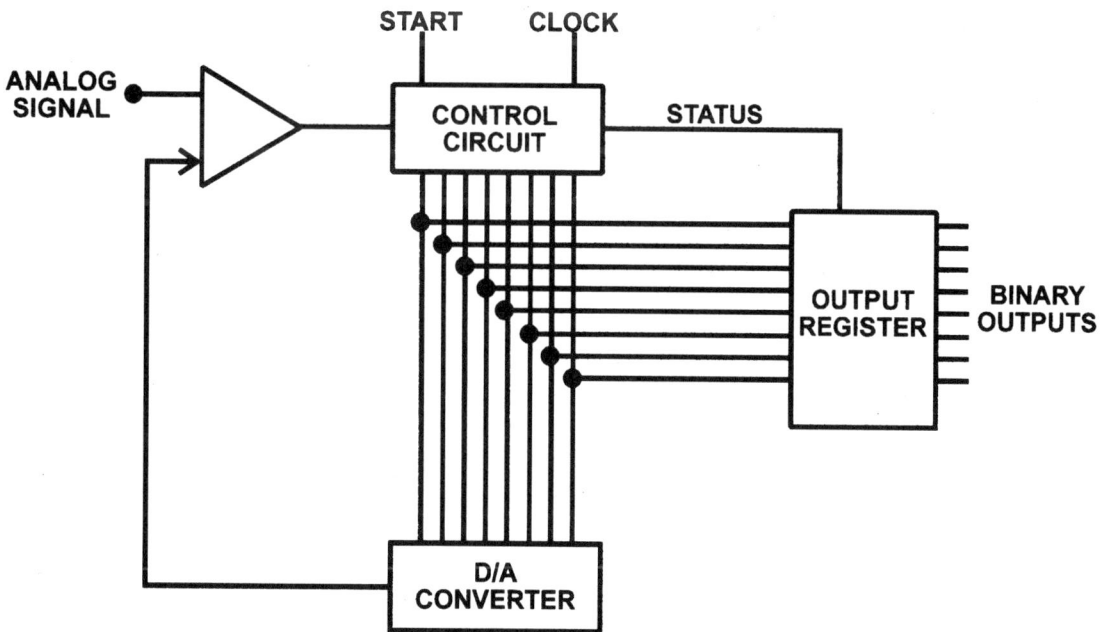

Figure 14-6. Successive approximation A/D converter.

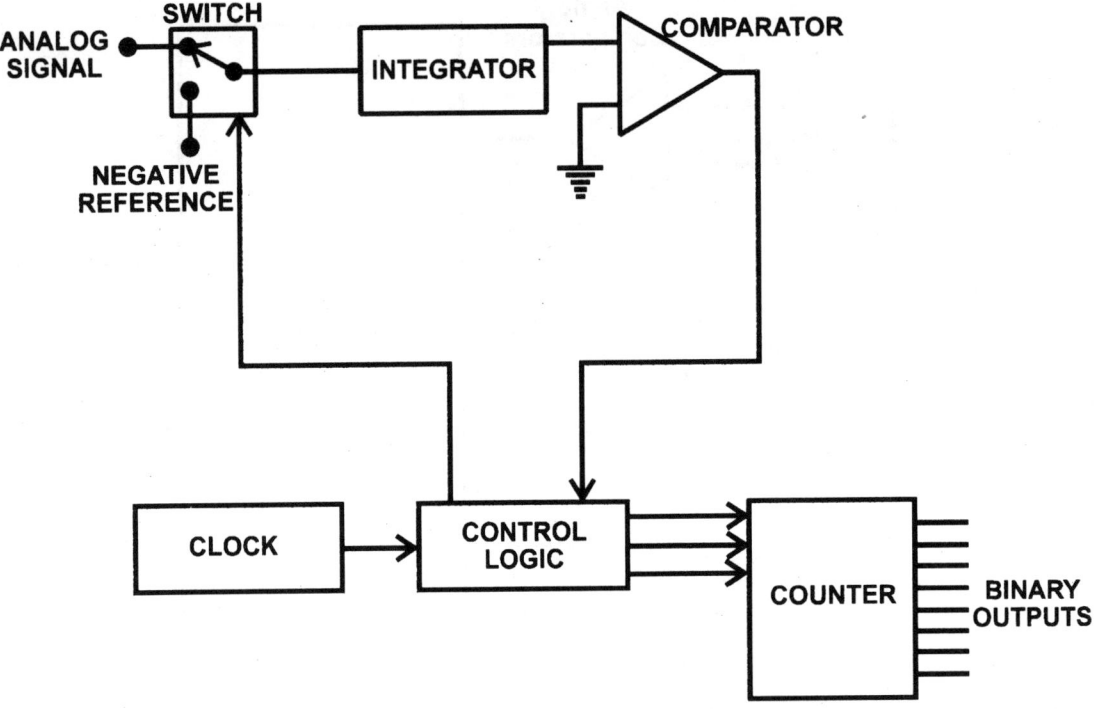

Figure 14-7. Dual slope A/D converter.

The conversion accuracy of a dual slope converter is independent of the clock frequency as long as the clock is stable and reasonably fast. Conversion accuracy is also independent of the accuracy of the integrator. The dual integration cancels any variations of the integrator. Dual slope converters are also simple, linear, and relatively low in cost. The main disadvantage of a dual slope converter is its long conversion time, which is usually several milliseconds.

Counter A/D Converter

The counter A/D converter is similar to a successive approximation converter. The main difference is in the control circuit. The D/A converter is driven by a counter. As the counter counts from 0000, 0001 to 1111, the converter successively sends 0, (1/16)FS to FS (full scale) to the comparator. When the reference input signal equals or exceeds the unknown signal, the conversion stops. The control circuit sends the counter output to the output register and turns on an EOC (end of conversion) signal to indicate that the binary output is available.

The counter A/D converter is simple, inexpensive, and accurate. It has a long conversion time because it is proportional to the input signal.

PROBLEMS

Problem 14-1.

Draw a block diagram of a digital thermometer.

Problem 14-2.

How does a digital-to-analog converter function?

Problem 14-3.

What is the output current of a binary weighted ladder D/A converter if R = 1000 ohms, Vref = 10 volts, and the applied binary word is 11010110?

Problem 14-4.

What is the disadvantage of a binary weighted ladder for binary words that are more than 8 bits long?

Problem 14-5.

What is the output voltage of a R-2R ladder converter if Vref = 10 volts and the input word is 11011011?

Problem 14-6.

What is the full scale output voltage of a R-2R ladder converter if Vref = 10 volts and the word is 8 bits long?

Problem 14-7.

What is the settling time of a D/A converter?

Problem 14-8.

How does an analog-to-digital converter function?

Problem 14-9.

Please compare parallel, successive approximation, counter, and dual slope A/D converters.

Problem 14-10.

How many comparators are needed for an 8 bit word in a parallel converter?

Problem 14-11.

What is the conversion time of a successive approximation A/D converter if the clock pulse is 10 MHz and the binary word is 8 bits long?

Problem 14-12.

What is the input signal to a dual slope A/D converter if the counter counts ten clock pulses, the holding number is two, and Vref = 10 volts?

Problem 14-13.

What is the main difference between a successive approximation A/D converter and a counter A/D converter?

Chapter Fifteen
THE FUTURE OF DIGITAL ELECTRONICS

Early computers used vacuum tubes as active devices. These computers filled a room, and they could only do simple arithmetic functions. Today, thanks to digital electronics, computers are much smaller, much faster, and can perform complex tasks. Digital electronics will probably continue to get smaller and faster. Here are a few predictions based on current developments in digital electronics. They may sound incredibly futuristic, but many of them are already becoming reality:

In the future, microminiature digital electronic circuits will be capable of performing incredible tasks. Micro-medical robots will be able to travel through our arteries and veins to dislodge and dissolve clots and to perform simple surgery. Micro-medical machines will be able to replace failing hearts. Rejection by our immune systems will be a thing of the past. Micro-medical robots will be able to deliver medicine to the site of infection. Chemotherapy will also be a thing of the past. The physical exam of the future will be quite different from what it is today. A micro-medical robot equipped to send back images to a computer will cruise our bodies in search of abnormalities and problems.

As computers become smaller and more capable, space flight vehicles will also become smaller and more capable. Humans will be able to travel longer distances, and smaller rocket engines will be required to take us there.

The home of the future will have a central computer for shopping. The grocery list will be entered on the computer and a few hours later the groceries will be delivered to the home. The computer will also pay for the groceries by debiting your bank account. This particular development may become commonplace much sooner than we think.

The lighting and heating requirements for the home can be preprogrammed on to the computer of tomorrow. If something is to be cooked in the oven while you are away, this too can be controlled by the computer of tomorrow. Again, this is already undergoing the process of becoming a reality.

The television of tomorrow will hang on the wall like a picture. The screen will employ liquid crystal display technology, and the electronics will be a micro-module mounted on the rear of the television LCD screen. The television of tomorrow will also have virtual reality capabilities. This technology has already been developed and is probably only a few years away from entering households en masse.

Automobiles of the future will have less wiring to perform more complex tasks than in the automobiles of today. Computers will be fast enough to send several signals simultaneously on a single pair of wires. Navigation systems will also pilot the car of tomorrow. The vehicle of the future will be equipped with radar systems to prevent accidents by applying the brakes or by steering as required. Windows will have liquid crystal display technology. The user will operate a switch if he or she wants the windows transparent for sunlight or opaque for privacy. These functions will also be controlled by a computer to save energy at night by making the windows opaque. Several automobile manufacturers are already working on these functions.

Education in the home is already being received by many people through their personal computers, and this practice will continue to grow. The computer will be able to teach as well as to administer tests to monitor the child's progress. Education will be more fun on the computer, too.

More and more people are working at home thanks to modern technology, and even more will be able to enjoy this reality in the near future. Their work will be sent via computers to the head office of their employers. Companies will only have one central office instead of dozens of offices across the nation; therefore, companies will be more efficient and more profitable.

Appendix
PROBLEM SOLUTIONS

CHAPTER ONE

Problem 1-1.

Switch: Digital.

Relay: Coil is analog and the contacts are digital.

Faucet: Analog.

Amplifier: Analog.

Oscillator: Analog.

Television: Analog.

Dimmer light switch: Analog.

Light bulb: Analog.

Computer mainframe: Digital.

Computer screen: Analog.

INPUT	OUTPUT
1	1
0	0

Table P1-1. Truth table for relay of *Problem 1-3*.

Problem 1-2.

Analog devices are neither positive nor negative logic. A switch is positive logic if the "ON" position applies power to the circuit. Relay contacts are positive logic if they are NORMALLY OPEN. "Normally open" or "normally closed" denotes the contact position when no power is applied to the relay coil. A computer may be positive or negative logic depending upon the digital circuits used in the computer.

Problem 1-3.

See *Table P1-1*.

CHAPTER TWO

Problem 2-1.
See *Table P2-1*.

Problem 2-2.
See *Table P2-2*.

Problem 2-3.
See *Table P2-3*.

x	y	$\overline{x+y}$
0	0	1
0	1	0
1	0	0
1	1	0

Table P2-1. Truth table for NOR gate.

x	y	$\overline{xy}$
0	0	1
0	1	1
1	0	1
1	1	0

Table P2-2. Truth table for NAND gate.

x	y	$\overline{x \oplus y}$
0	0	1
0	1	0
1	0	0
1	1	1

Table P2-3. Truth table for XOR gate with inverter.

Problem 2-4.
See *Table P2-4*.

Problem 2-5.

Let L = 0 when the lamp is OFF. Let L = 1 when the lamp is ON.

Let A = 0 when the switch is UP. Let A = 1 when the switch is DOWN.

Let B = 0 when the switch is UP. Let B = 1 when the switch is DOWN.

L = AB + $\overline{A}\overline{B}$. See *Table P2-5*.

x	y	z	f
0	0	0	0
0	0	1	0
0	1	0	0
0	1	1	1
1	0	0	0
1	0	1	1
1	1	0	0
1	1	1	1

Table P2-4. Truth table for *Problem 2-4*.

Problem 2-6.
f = WX + $\overline{X}\overline{Y}$ + $\overline{Z}$

CHAPTER THREE

Problem 3-1.

Fan-in is the number of inputs to a gate.

Fan-out is the number of identical gates that a logic gate can drive.

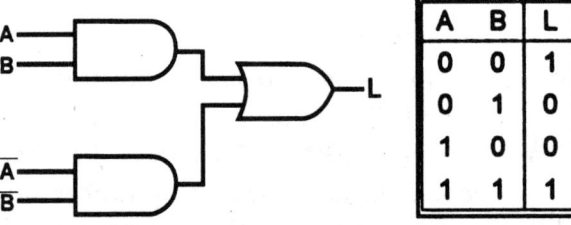

A	B	L
0	0	1
0	1	0
1	0	0
1	1	1

Table P2-5. Truth table and logic diagram for *Problem 2-5*.

Current sourcing is when current flows from the gate output to the load.

Appendix / 113

Current sinking is when current flows from the load of a gate to the output of that gate.

Noise margin is the limit in tolerance spread and the degree of loading of a gate.

Problem 3-2.

Propagation delay is the time it takes a signal to travel through a gate.

Problem 3-3.

RTL.

Problem 3-4.

"0"NM = $V_{iL} - V_{oL}$ = 2 - 1 = 1 volt.

"1"NM = $V_{oH} - V_{iH}$ = 10 - 8 = 2 volts.

Problem 3-5.

ECL.

Problem 3-6.

CMOS because of its low power consumption.

Problem 3-7.

See *Table P3-1*.

	RTL	DTL	TTL	CMOS	ECL
Fan-in	4 or less	4 or less	8		
Type	Current sourcing	Current sinking			
Noise Immunity	Fair	Good	Very Good	Good	Good
Fan-out	Poor	Good	Good ≈12	Good	Good ≈25
Speed	Poor for low power, ≈30nS Good for high power, ≈12nS	Good 25 - 80nS	Best for saturated circuits, 13nS	High speed	High speed

Table P3-1. Comparison of logic families.

CHAPTER FOUR

Problem 4-1.

A.0 = 0, Th. (a) and A.A = A, Th. (c).

A.1 = A, Th. (b) and A.$\overline{A}$ = 0, Th. (d).

A.0 = 0, Th. (a) and A.A = 0, Th. (d).

A.1 = A, Th. (b) and A.A = A, Th. (c).

See *Table P4-1* for the AND truth table.

Problem 4-2.

A + 0 = 0, Th. (e) and A + A' = A, Th. (g).

A + 1 = 1, Th. (f) and A + $\overline{A}$' = 1, Th. (h).

A + 0 = A, Th. (e) and A + A = 1, Th. (h).

A + 1 = 1, Th. (f) and A + A = A, Th. (g).

See *Table P4-2* for the OR truth table.

Problem 4-3.

Th. (n) states that AB + AB' = A. The truth table and logic diagram for Th. (n) appear under *Table P4-3*.

Problem 4-4.

See *Table P4-4*.

Problem 4-5.

See *Table P4-5*.

Problem 4-6.

See *Table P4-6*.

Problem 4-7.

See *Table P4-7*.

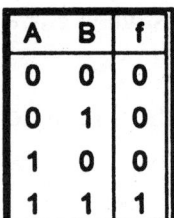

Table P4-1. Truth table for AND gate.   *Table P4-2.* Truth table for OR gate.

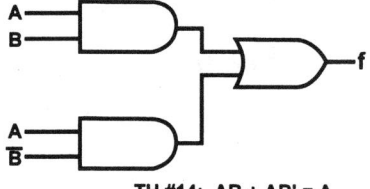

TH.#14: AB + AB' = A

A	B	f	
0	0	0	
0	1	0	f = A
1	0	1	
1	1	1	

Table P4-3. Truth table and logic diagram for Theorem #14.

Table P4-4. Karnaugh map for 2-input AND gate.

Table P4-5. Karnaugh map for 3-input OR gate.

Table P4-6. Karnaugh map for 4-input AND gate.

Table P4-7. Karnaugh map for a four-variable function.

Problem 4-8.

See *Table P4-8*.

Problem 4-9.

See *Figure P4-1*.

Problem 4-10.

See *Figure P4-2*.

Problem 4-11.

See *Figure P4-3*.

Problem 4-12.

See *Table P4-9*.

Problem 4-13.

See *Figure P4-4*.

		0	0	0	0	1	1	1	1	A
		0	0	1	1	1	1	0	0	B
		0	1	1	0	0	1	1	0	C
0	0	f_0	f_4	f_{12}	f_8	f_{24}	f_{28}	f_{20}	f_{16}	
0	1	f_1	f_5	f_{13}	f_9	f_{25}	f_{29}	f_{21}	f_{17}	
1	1	f_3	f_7	f_{15}	f_{11}	f_{27}	f_{31}	f_{23}	f_{19}	
1	0	f_2	f_6	f_{14}	f_{10}	f_{26}	f_{30}	f_{22}	f_{18}	
D	E									

Table P4-8. Karnaugh map for a five-variable function.

CHAPTER FIVE

Problem 5-1.

See *Figure P5-1*. If both inputs are at the enabled level when the circuit is clocked, the output is uncertain.

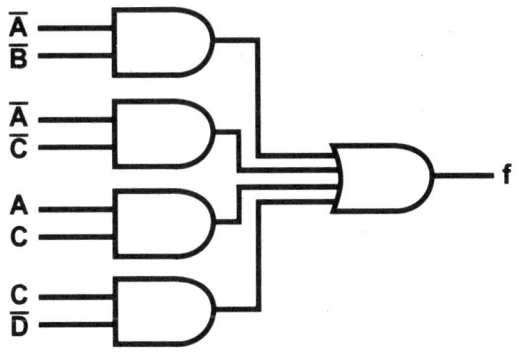

Figure P4-1. Logic circuit for *Problem 4-9*.

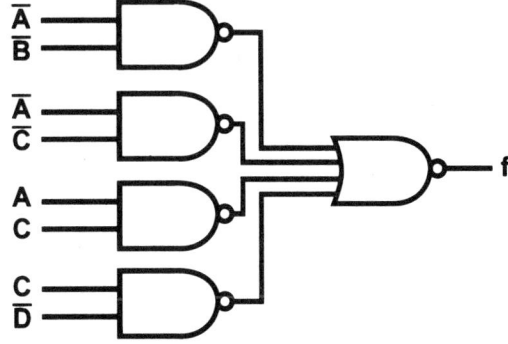

Figure P4-2. Logic circuit for *Problem 4-10*.

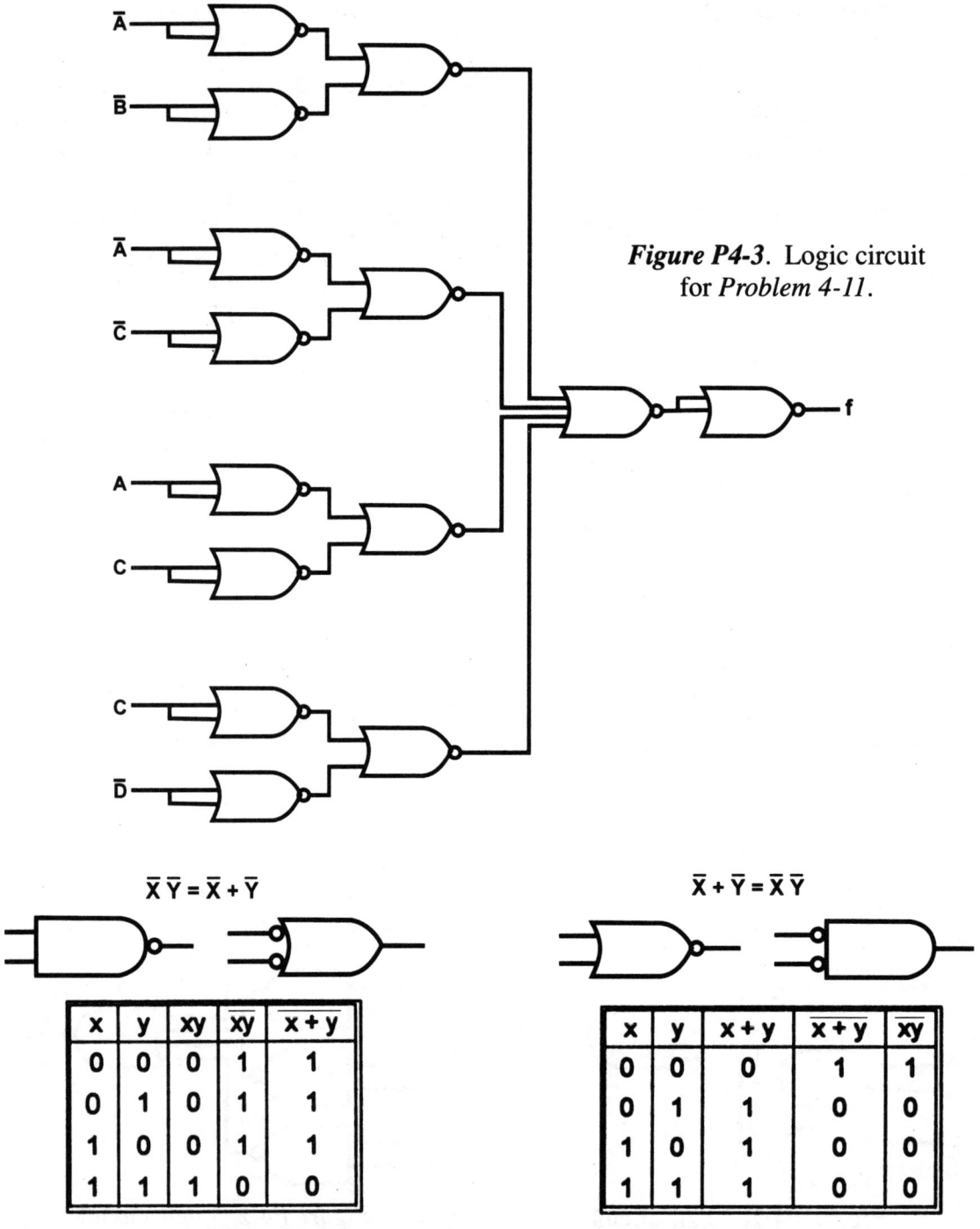

Figure P4-3. Logic circuit for *Problem 4-11*.

Table P4-9. Truth tables for DeMorgan's theorems.

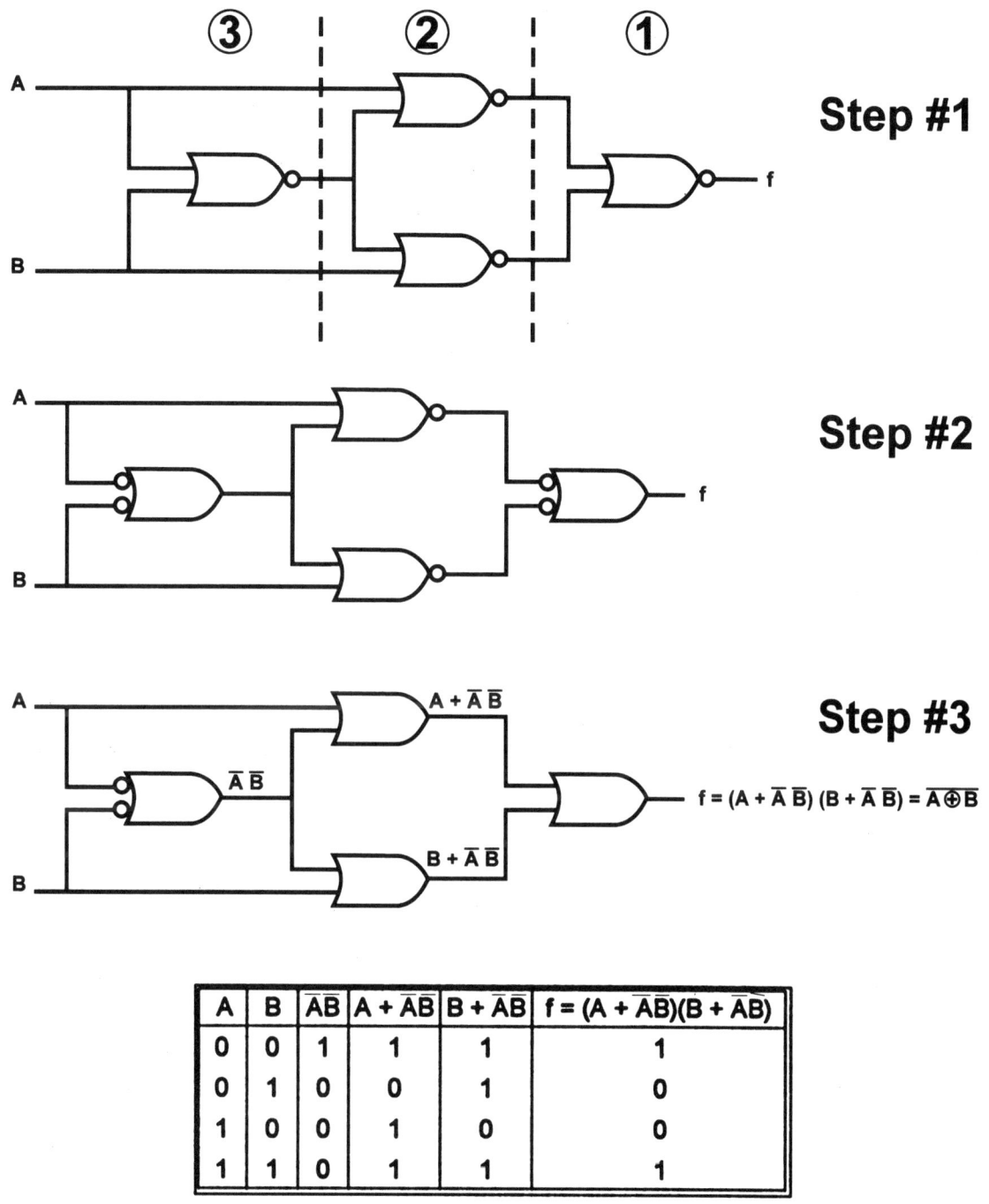

Figure P4-4. Simplified logic circuit for *Problem 4-13*.

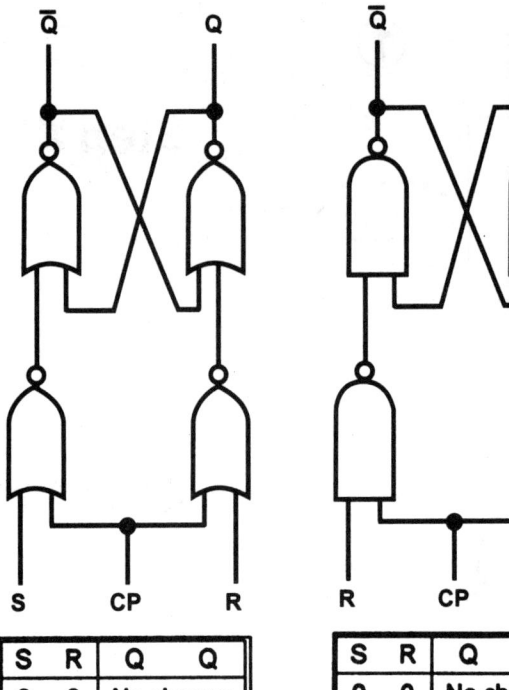

RS		Q (After clocking)
0	0	No change
0	1	1
1	0	0
1	1	Undefined

Table P5-1. Checked RS flip-flop truth table

Figure P5-1. Clocked and gated flip-flop using NOR gates (a) and NAND gates (b).

Problem 5-2.
See *Figure P5-2.*

Problem 5-3.
Direct set and *direct clear* can be used to set or reset the flip-flop independent of the clock pulse. This is achieved in a master-slave flip-flop by acting directly upon the master and slave latches.

Problem 5-4.
See *Table P5-1.*

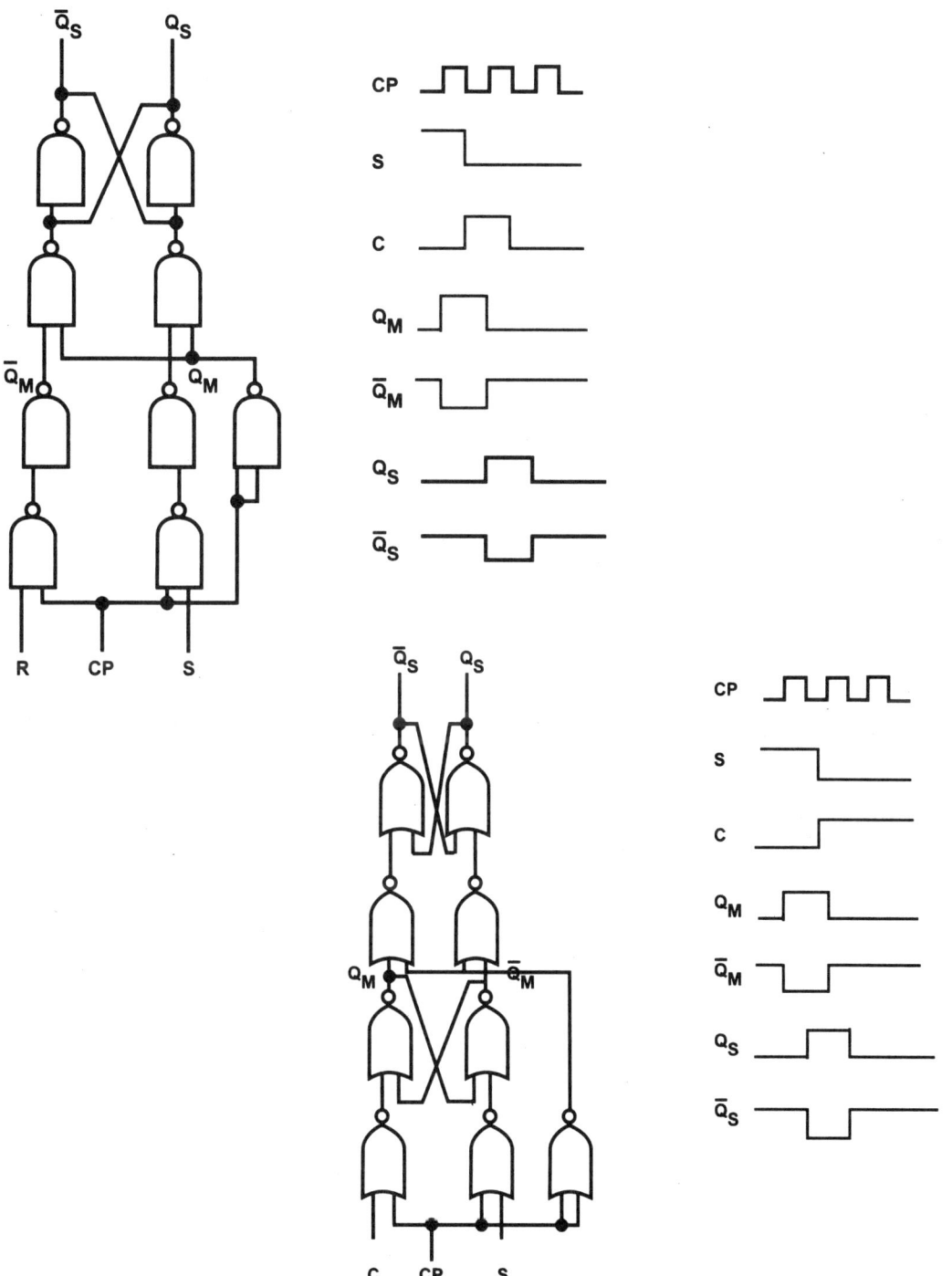

Figure P5-2. Master-slave flip-flop using NAND gates (a) and NOR gates (b).

Figure P5-3. Timing diagram for *Problem 5-5.*

Problem 5-5.

See *Figure P5-3*. At the trailing edge of each clock pulse, each flip-flop accepts the state of the previous flip-flop.

Problem 5-6.

See *Figure P5-4*.

Problem 5-7.

See *Figure P5-5*. The JK flip-flop can have both J and K "high." This is not possible in a clocked RS flip-flop.

CHAPTER SIX

Problem 6-1.

See *Table P6-1*.

Problem 6-2.

See *Figure P6-1*.

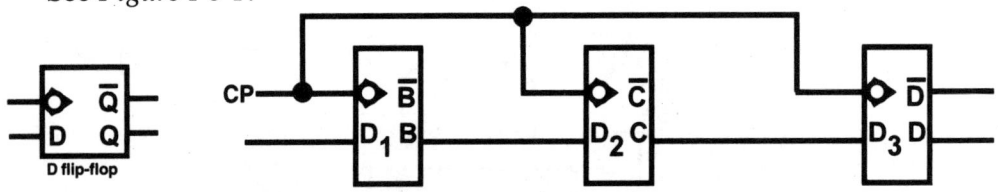

Figure P5-4. Logic circuit of a D flip-flop (a) and Figure 5-12 redrawn using D flip-flops (b).

Appendix / 121

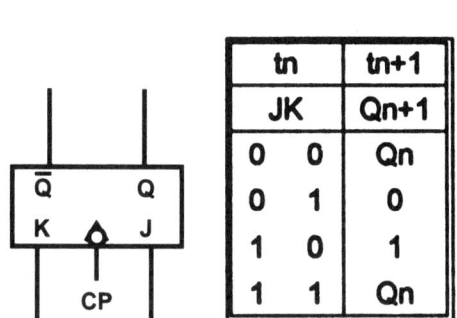

Figure P5-5. JK flip-flop logic circuit and truth table.

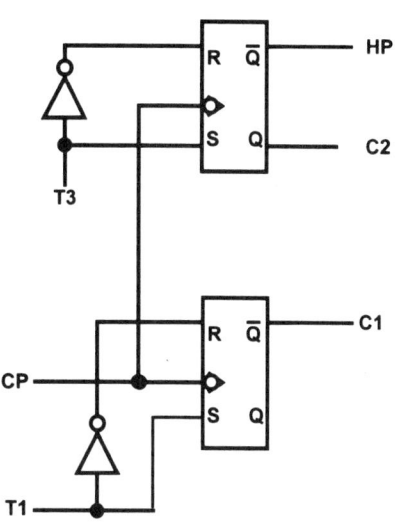

Figure P6-1. Sequential control circuit using RS flip-flop.

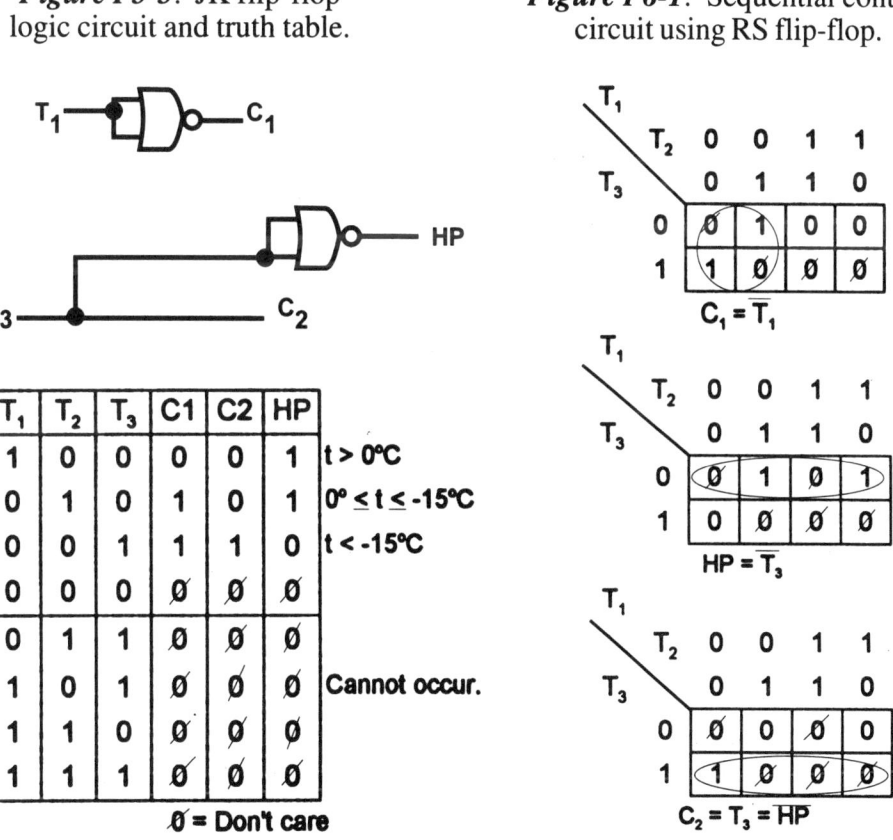

Table P6-1. NOR logic control circuit for *Problem 6-1*.

CHAPTER SEVEN

Problem 7-1.

$101.375 = 10^2 + 10^0 + 3 \times 10^{-1} + 7 \times 10^{-2} + 5 \times 10^{-3}$

$(= 100 + 1 + .300 + .070 + .005)$

Problem 7-2.

$101.375 = 2^6 + 2^5 + 2^2 + 2^0 + 2^{-2} + 2^{-3}$

$(= 64 + 32 + 4 + 1 + 1/4 + 1/8)$

Problem 7-3.

110011.011

Problem 7-4.

$59 = 59/8 = 7$, with a remainder of 3; therefore, $59_8 = 7\ 3$

Problem 7-5.

```
            10110
    11001 ) 1000100110
            11001
            00100101
               11001
               00011001
                  11001
                  00000
```

Problem 7-6.

See *Table P7-1.*

Problem 7-7.

See *Table P7-2.*

Problem 7-8.

See *Table P7-3.*

0	0	0	0
0	0	0	1
0	0	1	0
0	0	1	1
0	1	0	0
0	1	0	1
0	1	1	0
0	1	1	1
1	0	0	0
1	0	0	1

Table P7-1. BCD code.

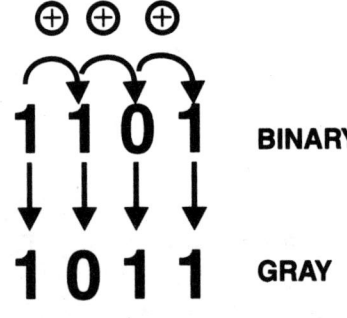

Table P7-2. Converting a binary bit to Gray code.

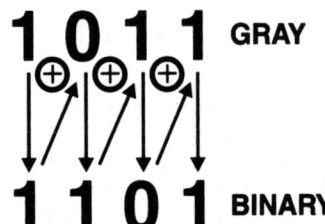

Table P7-3. Converting a Gray bit to binary code.

CHAPTER EIGHT

Problem 8-1.
A *register* is an assembly of flip-flops capable of storing binary data.

Problem 8-2.
Parallel entry is when all of the bits of a binary word are transferred at the same time.

Problem 8-3.
Asynchronous operation may set or clear flip-flops directly at any time. It saves time by using the time between clock pulses.

Problem 8-4.
See *Table P8-1*.

Problem 8-5.
In serial data entry, the flip-flops must be master-slave or edge-triggered so that a data bit at the input will not race through the register when the clock pulse is high.

Problem 8-6.
See *Figure P8-1*.

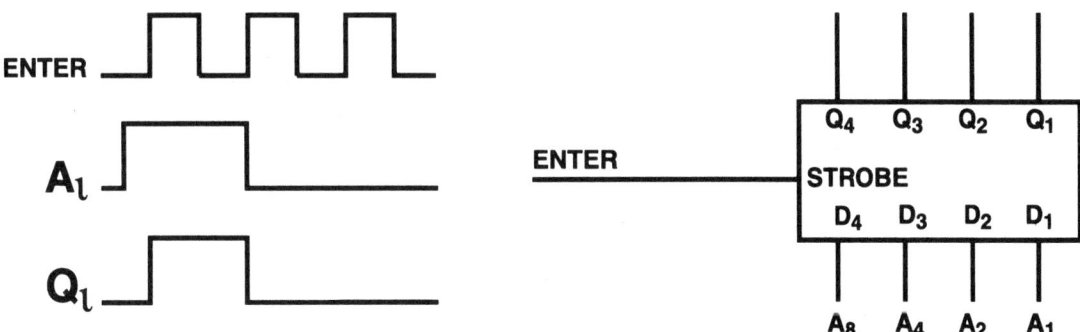

Table P8-1. Timing diagram for entry into gated latch register.

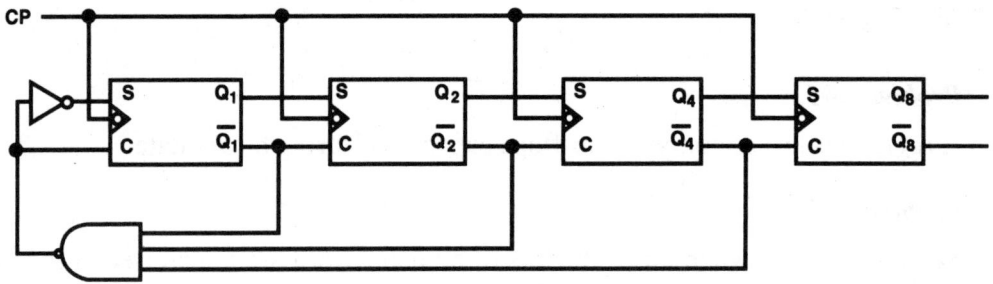

Figure P8-1. Self-starting counter.

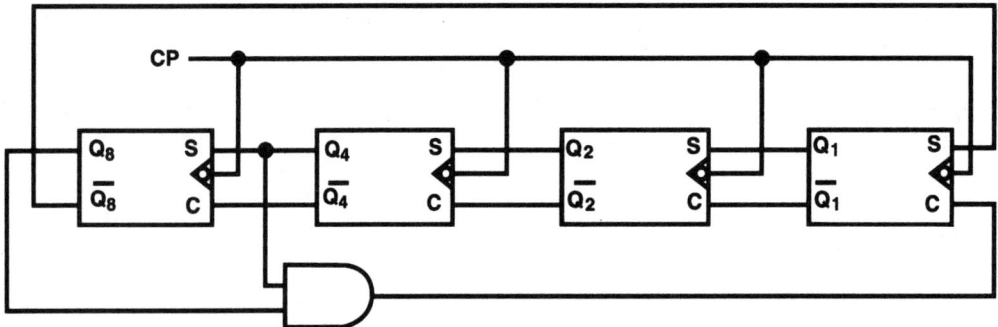

Figure P8-2. Self-starting complementing ring counter.

Problem 8-7.

See *Figure P8-2*.

Problem 8-8.

See *Table P8-2*.

CHAPTER NINE

Problem 9-1.

Encoders change many lines into a few lines.

Decoders change a few lines into many lines.

Multiplexers are data selectors.

Problem 9-2.

See *Table P9-1*.

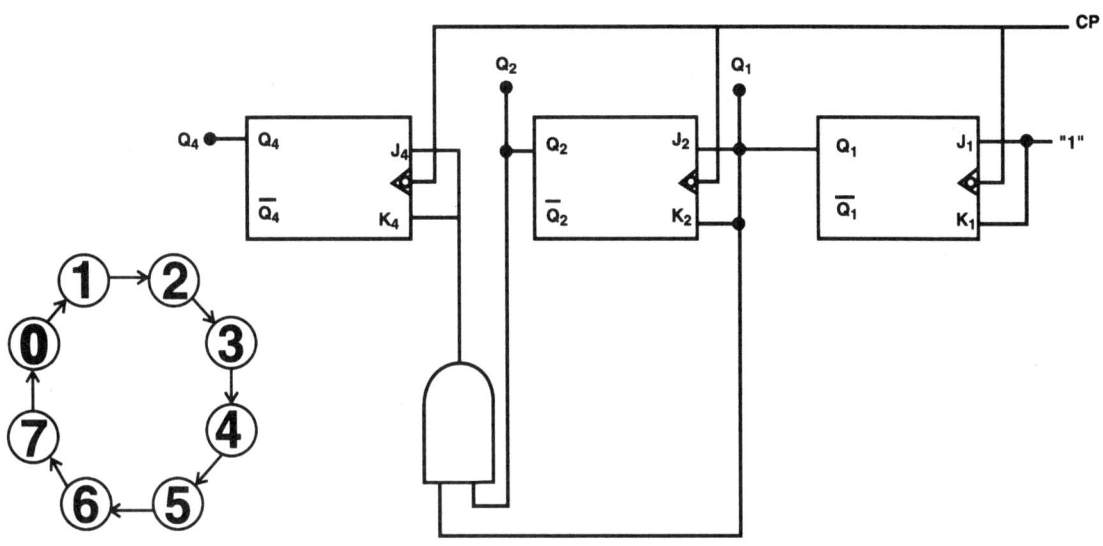

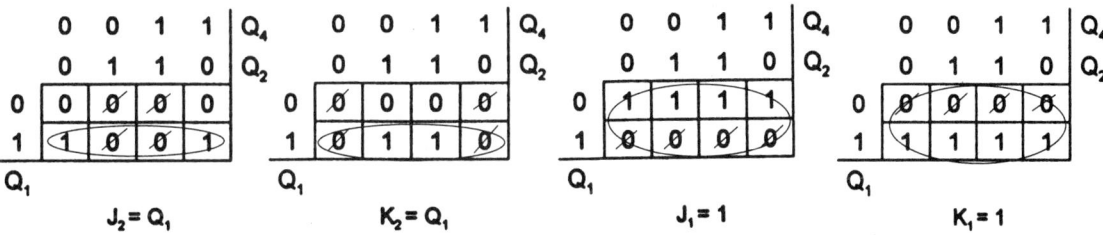

Table P8-2. Binary counter design.

	Rectangular decoder	Tree decoder	Dual-tree decoder
One AND output HIGH at any time	—	—	
n flip-flops require 2^n n-input AND gates	—	n flip-flops require $2 \times 2^{n/2}$ n/2-input AND gates	
Fast	Slower	Slowest	
Most connections	Less connections	Least connections	
Most input connections	Less input connections	Least input connections	

Table P9-1. Comparison of rectangular, tree, and dual tree decoders.

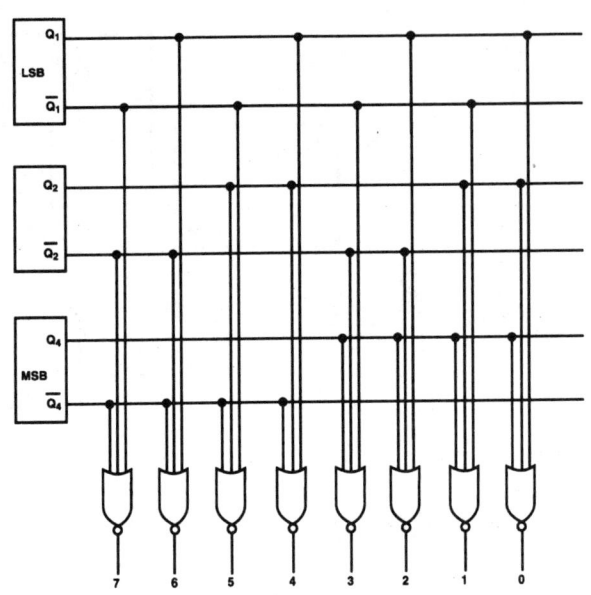

Figure P9-1. Circuit for *Problem 9-3*.

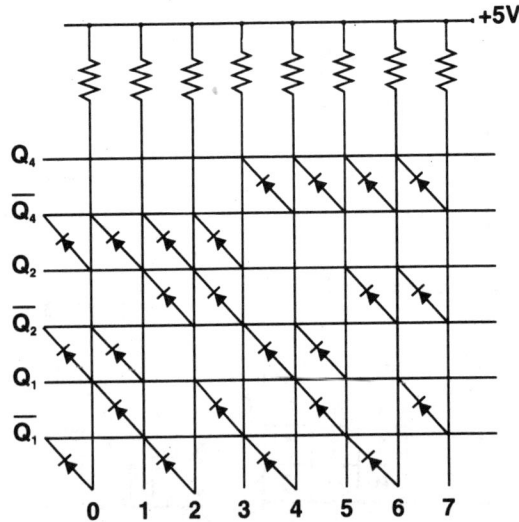

Figure P9-2. Rectangular decoder using diodes.

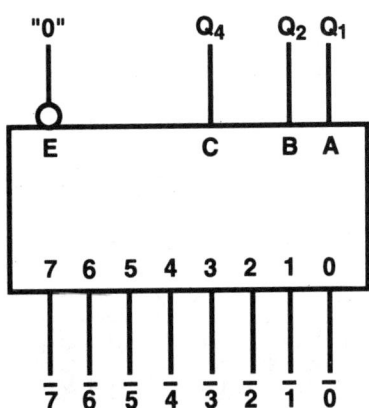

Figure P9-3. 1/8 decoder module.

Problem 9-3.

See *Figure P9-1*.

Problem 9-4.

See *Figure P9-2*.

Problem 9-5.

Let n = #. Flip-flops = 6.

Rectangular decoder:

2^n n-input AND gates, 2^6 6-input AND gates; that is, 64 6-input AND gates.

Total: 384 inputs.

Tree decoder:

1st level: n 2-input AND gates = 6 2-input AND gates.

2nd level: 2n 2-input AND gates = 12 2-input AND gates.

3rd level: 3n 2-input AND gates = 18 2-input AND gates.

4th level: 4n 2-input AND gates = 24 2-input AND gates.

5th level: 5n 2-input AND gates = 30 2-input AND gates.

Total: 90 2-input AND gates = 180 inputs.

Dual tree decoder:

Split into two, let n1 = n2 = 3.

2^n n-input AND gates = 8 3-input AND gates for each half.

Total: Twice 8 3-input AND gates = 16 3-input AND gates = 48 inputs.

Problem 9-6.

See *Figure P9-3*.

Problem 9-7.

See *Figure P9-4*.

Problem 9-8.

See *Figure P9-5*.

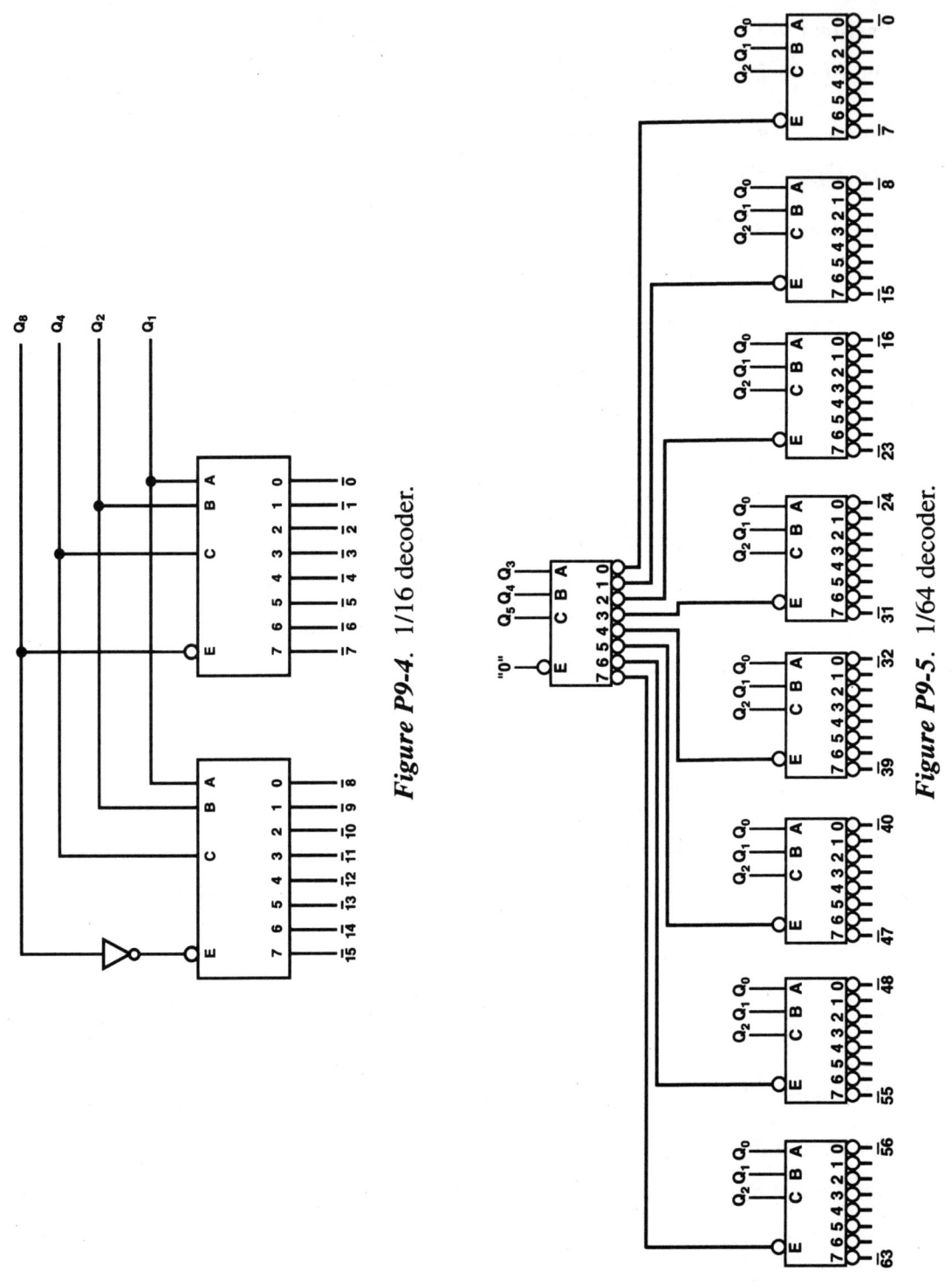

Figure P9-4. 1/16 decoder.

Figure P9-5. 1/64 decoder.

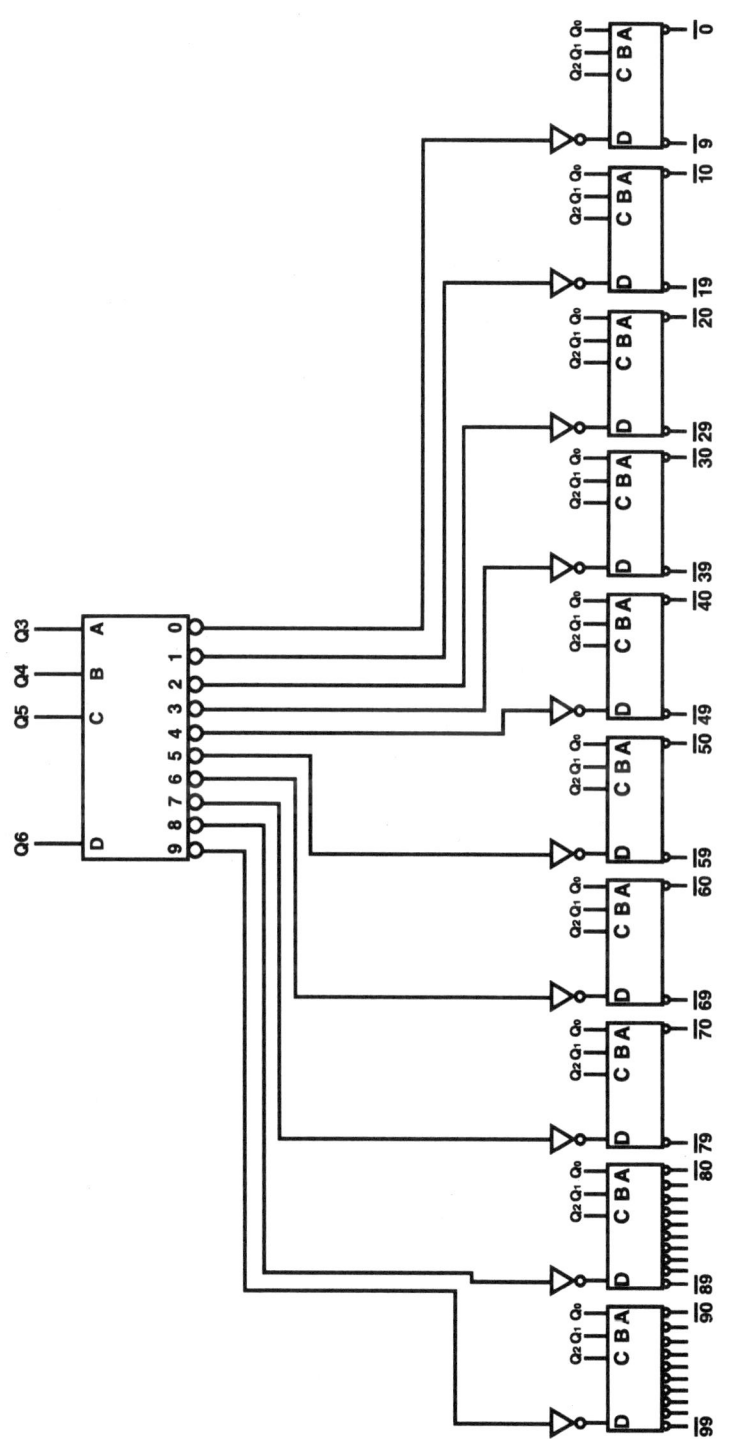

Figure P9-6. 1/100 decoder.

130 / Digital Electronics

Problem 9-9.

Requires eleven 1/10 decoders (similar to *Figure P9-5*), as shown in *Figure P9-6*.

Problem 9-10.

See *Figure P9-7*.

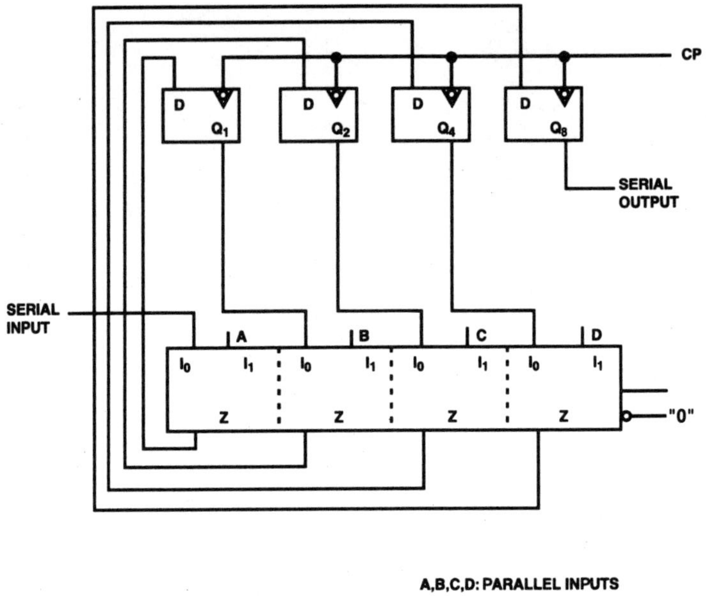

CHAPTER TEN

Problem 10-1.

See *Table P10-1*.

Problem 10-2.

See *Figure P10-1*.

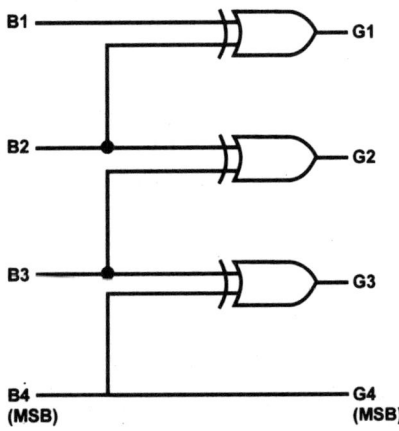

Figure P9-7. Register mode control.

Problem 10-3.

The (A>B), (A<B) and (A = B) inputs feed corresponding outputs of the IC, but are conditioned by the equality of each bit pair on the IC. This provides the means of including on the IC the results determined by the lower bit positions. See *Figure P10-2*.

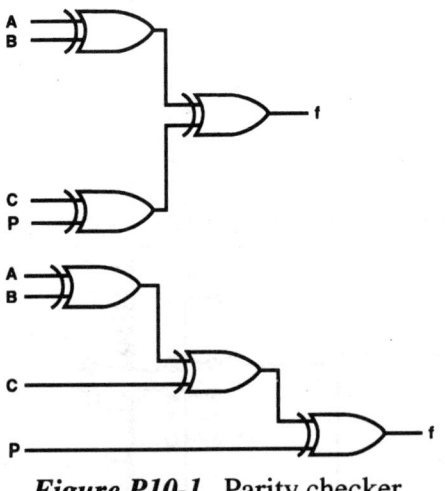

Figure P10-1. Parity checker.

Figure P10-3. Binary code to Gray code converter.

A	B	C	D	f⊕
0	0	0	0	0
0	0	0	1	1
0	0	1	0	1
0	0	1	1	1
0	1	0	0	1
0	1	0	1	1
0	1	1	0	1
0	1	1	1	1
1	0	0	0	1
1	0	0	1	1
1	0	1	0	1
1	0	1	1	1
1	1	0	0	1
1	1	0	1	1
1	1	1	0	1
1	1	1	1	0

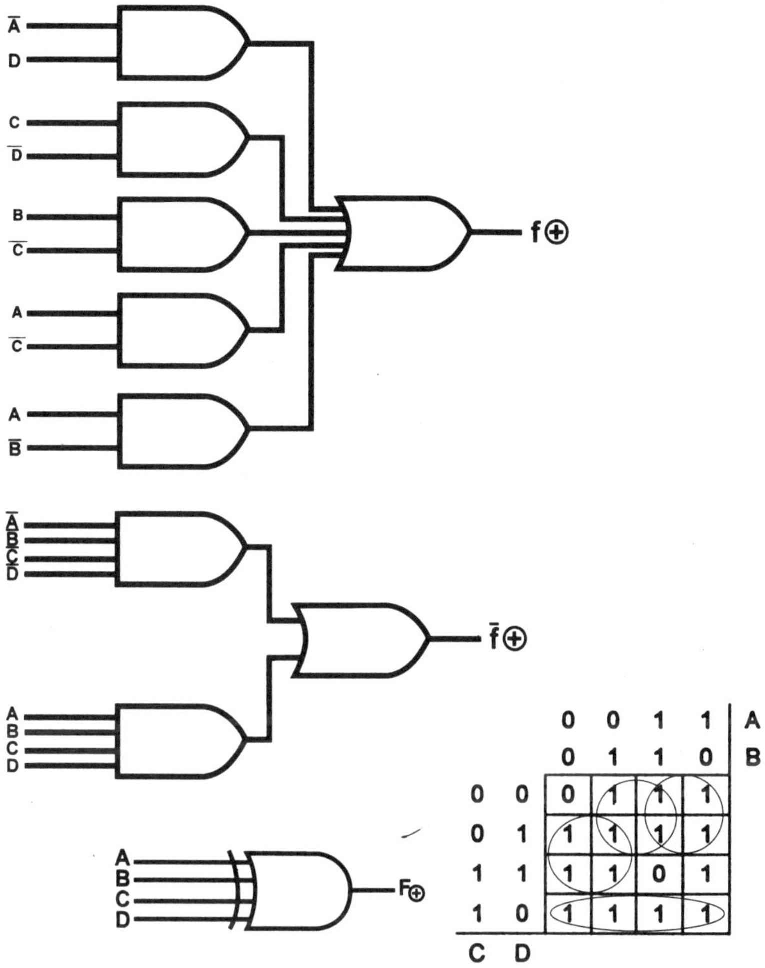

$f⊕ = \overline{A}D + A\overline{C} + C\overline{D} + \overline{C}B + \overline{A}B$

$\overline{f}⊕ = \overline{ABCD} + ABCD$

Table P10-1. Karnaugh maps and circuits for *Problem 10-1*.

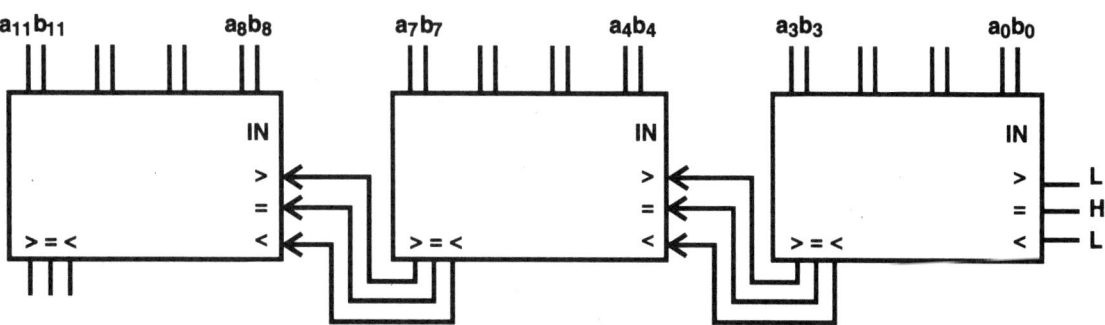

Figure P10-2. LSB first relative magnitude detector.

Problem 10-4.
See *Figure P10-3*.

CHAPTER ELEVEN

Problem 11-1.
Let n = # flip-flops for the binary counter. Therefore, n = 3.

Problem 11-2.
In a ripple counter, each flip-flop toggles on the trailing edge of the clock pulse.

Problem 11-3.
Ripple counters with many stages may have a propagation delay exceeding the clock pulse period. The higher order flip-flops may not change states accordingly.

Problem 11-4.
Ripple counters are used where only the output of the last flip-flop is used. Therefore, ripple transients do not affect the application. A frequency divider is a typical ripple counter application.

Problem 11-5.
All of the flip-flops of a synchronous counter are clocked simultaneously by the same clock pulse. Flip-flops change states simultaneously with the same propagation delay.

Problem 11-6.
Pseudosynchronous operation is when the flip-flops do not change states simultaneously. The change of state is controlled by the J and K inputs.

Problem 11-7.
A pseudosynchronous counter is a synchronous counter when all of its gate delays are the same.

Problem 11-8.
A serial carry counter has gates in series or cascaded with previous J and K inputs and with previous Q outputs. The maximum clock rate is less than that of a parallel carry counter because JK input gates are cascaded.

Appendix / 133

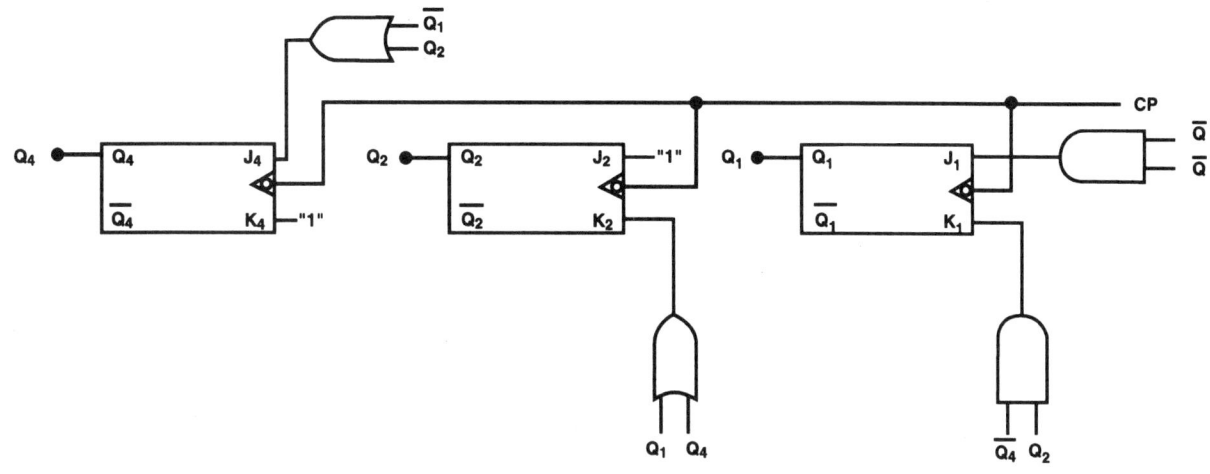

Table P11-1. Counter design for *Problem 11-10*.

Present									Next		
Q_4	Q_2	Q_1	$J_4=\bar{Q}_1+Q_2$	$K_4=1$	$J_2=1$	$K_2=Q_1+Q_4$	$J_1=\bar{Q}_4\bar{Q}_2$	$K_1=\bar{Q}_4Q_2$	Q_4	Q_2	Q_1
0	0	0	1	1	1	0	1	0	1	1	1
0	0	1	0	1	1	1	1	0	0	1	1
0	1	0	1	1	1	0	0	1	1	1	0
0	1	1	1	1	1	1	0	1	1	0	0
1	0	0	1	1	1	1	0	0	0	1	0
1	0	1	0	1	1	1	0	0	0	1	1
1	1	0	1	1	1	1	0	0	0	0	0
1	1	1	1	1	1	1	0	0	0	0	1

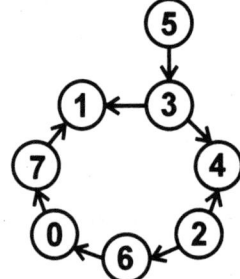

Table P11-2. Verification of counter design of *Problem 11-10*.

Problem 11-9.

A parallel carry counter has gates in parallel with previous J and K inputs and previous Q outputs.

Problem 11-10.

See *Table P11-1*.

Problem 11-11.

See *Table P11-2*.

CHAPTER TWELVE

Problem 12-1.

The half adder does not consider previous sums, while the full adder takes previous carries into account.

Problem 12-2.

The worst-case addition time of a ripple adder occurs when 0001 and 1111 are added together because each step generates a carry.

Problem 12-3.

The ripple adder accumulator must be made with synchronous flip-flops or registers.

Problem 12-4.

The look-ahead adder does not wait for a carry to ripple through it like a ripple adder.

Problem 12-5.

The look-ahead adder examines the inputs of all previous stages to determine what its input carry will be.

Problem 12-6.

The slowest application of a ripple adder is a character-to-character ripple adder.

Problem 12-7.

In the fast adder, the carry generated in or propagated through the least significant section is passed directly to the second carry input of the most significant section, if the propagation functions of the other stages are "ONE."

Problem 12-8.

See *Figure P12-1*.

Problem 12-9.

```
  011
 x101
  011
 0000
 01100
 01111
```

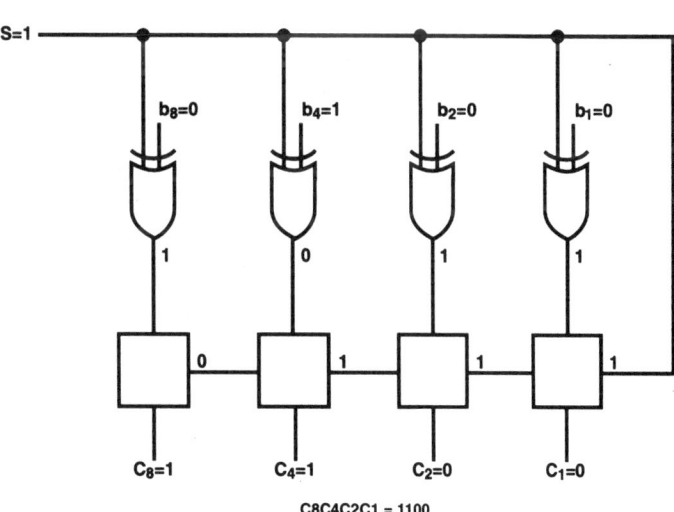

Figure P12-1. Circuit for *Problem 12-8*.

CHAPTER THIRTEEN

Problem 13-1.

Internal memory is an integral part of the system. It does not need input/output procedures to transfer data. External memory is not an integral part of the system. It needs input/output procedures to transfer data. External memory is slower than internal memory.

Problem 13-2.

Let n = number of address inputs. The number of addresses = $2^n = 2^6 = 64$.

Problem 13-3.

Access time is the time it takes for data to be available once a request for data is recognized.

Memory cycle time is the minimum interval between the reading of two words.

Problem 13-4.

ROM stores data permanently; rewriting is not possible.

RAM can have address selection made in any order.

Problem 13-5.

PROM is a ROM memory that is programmable by the manufacturer. It can only read memory after it is programmed by the manufacturer.

Problem 13-6.

EAROM is a ROM that can be erased and reprogrammed through a special process.

Problem 13-7.

Word-oriented memory is two-dimensional, has a sense line for reading, and a write line for writing. It is fast, and the decoder required is large and rather expensive.

Bit-oriented memory is three-dimensional, and needs two signals to select any bit. Only two small decoders are required per text.

Problem 13-8.

Let n = 5. X select lines = $2^{n/2} = 2^3 = 8$. Y select lines = $2^{n/2} = 2^3 = 8$.

Problem 13-9.

Type is bit-oriented: Let n = 5. Number of decoders = $2 \times 2^{n/2} = 2 \times 2^{2.5} = 2 \times 2^3 = 2 \times 8 = 16$.

Problem 13-10.

Type is word-oriented: Let n = 5. Number of decoders = $2^n = 2^5 = 32$. Hence, 1 out of 32 decoders is required.

Problem 13-11.

See *Figure P13-1*.

Problem 13-12.

See *Figure P13-2*.

Problem 13-13.

See *Figure P13-3*.

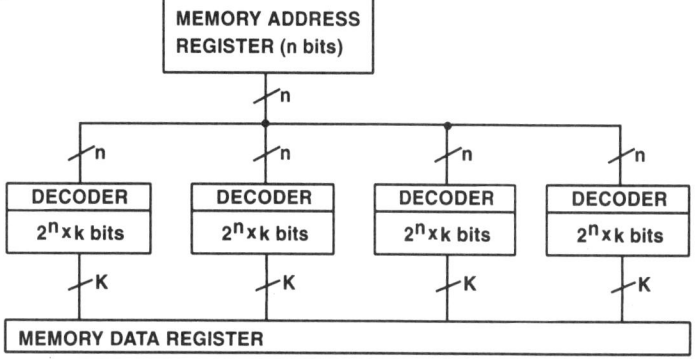

Figure P13-1. Expanding the word size of a memory.

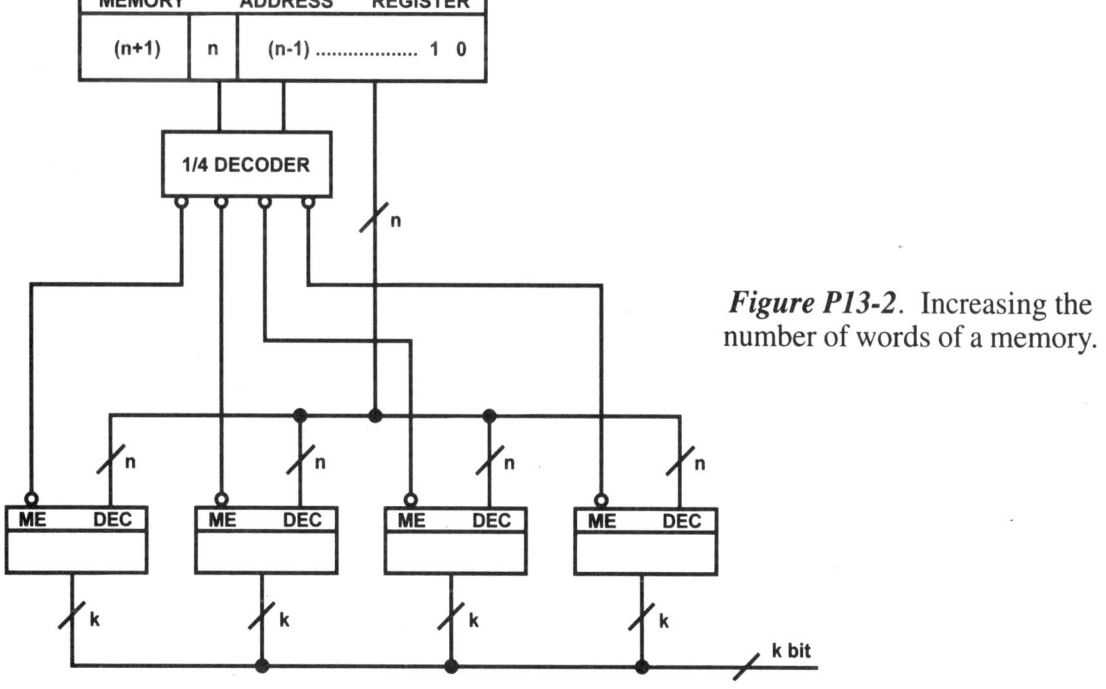

Figure P13-2. Increasing the number of words of a memory.

Problem 14-9.

See *Table P14-1*.

Problem 14-10.

Let n = 8. A parallel converter needs $2^n - 1$ comparators = $2^8 - 1$ = 255 comparators.

Problem 14-11.

Conversion time = $nT = n/f = 8/10 \times 10^{-6}$ = 0.8 usec.

Problem 14-12.

Vin = Vref(T2/T1) = 2/10 x 10 = 2 volts.

Problem 14-13.

The main difference between successive approximation and counter analog-to-digital converters is in the control circuit.

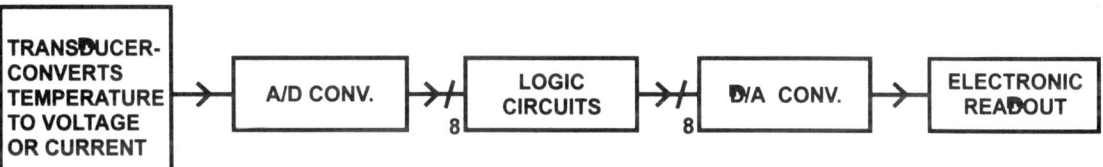

Figure P14-1. Block diagram of a digital thermostat.

Parallel	Successive approximation	Counter	Dual slope
Fast	High speed	Simple	Simple
		Inexpensive	Linear
		Accurate	Low in cost
Needs many comparators	High resolution	Long conversion time	Long conversion time
	Accurate		Accuracy independent of integrator accuracy and of clock frequency

Table P14-1. Comparison of analog-to-digital converters.

GLOSSARY

8421 code: Weighted code in which the sum of the weights of each group of four bits or words that are 1 equals the decimal digit that they represent.

Access: Scheme for making data available.

Access time: The time it takes for data to be available once the request for data is recognized.

Address: The location in memory of a word, byte or bit.

Analog circuit/signal: Varies continuously due to temperature, pressure, or speed transducers. "Real world" signals.

Analog-to-digital (A/D) converter: Converts an analog signal to a digital signal.

AND gate: Has the output high only when all of its inputs are high.

ASCII code: American Standard Code for Information Interchange. Used extensively in data transmission, in which 128 numerals, letters, symbols, and special codes are each represented by a 7-bit binary number.

Asynchronous counter: Lacks a regular time relationship, so is unpredictable with respect to instruction sequences.

Asynchronous inputs/outputs: Accepts input data while simultaneously producing output data.

Asynchronous operation: Operation independent of the clock pulse.

Binary arithmetic: Uses a base number of 2. Therefore, there are two digits that are used; 1 and 0.

Binary-coded decimal: System of number and character representation in which specific decimal digits are represented in binary code.

Biquinary code: Mixed-radix notation in which each decimal digit is considered to be the sum of two digit; 0 or 1 with significance 5, and 0, 1, 2, 3 or 4 with significance 1.

Bit-oriented: Also known as coincident signal select. Anything used to store or convey information using a unit equal to one binary decision. A bit is represented by either a 1 or a 0, and can also mean "yes" or "no."

Boolean algebra: Algebraic rules for manipulating logic equations. Deals with classes, propositions, on-off circuit elements, etc., presented as AND, OR, NOT, etc., allowing for computations and mathematical demonstrations.

Comparator: Has a high output when all of the inputs are the same.

Control circuit: Turns a device on or off in response to a specific set of input conditions.

Counter: A circuit with a specific count sequence.

Counter A/D converter: Provides a digital output after receiving specific analog inputs.

Current sinking: When current flows from the load to the driving circuit.

Current sourcing: When current flows from the driving circuit to the load.

D flip-flop: Output is determined by input that occurred one pulse earlier.

Decoder: Changes a few lines to many lines.

Demultiplexer: Takes data of a single input and distributes it to 2^n data lines.

Digital circuit: Operates like an "on/off" switch and can make logical decisions. Types include RTL, DTL, TTL, etc.

Digital code: A binary representation of data.

Digital-to-analog (D/A) converter: Converts a digital signal into an analog signal, or to a voltage or current with a magnitude proportional to the numeric value of the digital signal.

Diode transistor logic (DTL): Uses diodes at the input to the perform electronic logic function that activates the circuit transistor output.

Dual slope A/D converter: Converts an unknown signal into a proportional time interval which is in turn measured digitally.

Dumping: Reading the contents of an entire computer memory.

Dynamic memory: Semiconductor memory in which the presence or absence of an electrical charge represents the two states of a storage element.

Erasable and reprogrammable memory (EAROM): A ROM that can be erased and reprogrammed through a special process.

Encoder: Changes many lines to a few lines.

Equality detector: Outputs a 1 when its inputs are equal.

Excess-3 code: Based on adding 3 to a decimal digit then converting the result into binary form.

Exclusive-OR gate: A gate that is low when the inputs are the same.

External memory: Auxiliary storage unit outside of the internal memory of a computer.

Fan-in: The number of inputs to a gate.

Fan-out: The number of identical gates that a logic gate can drive.

Flip-flop: A logic circuit that can store an output.

Full-adder: Adds two bits to obtain a sum and a carry. It does take into account previous carries.

Gray code: Composed of a number of bits assigned so that only one bit changes at each increment or decrement.

Half-adder: Adds two bits to obtain a sum and a carry. It does not take into account previous carries.

Hamming code: Error correction system used in data transmission.

Inequality detector: Outputs a 1 when its inputs are different.

Internal memory: Total storage in a computer automatically available to the computer.

Inverter (NOT gate): Has the opposite output of the input.

JK flip-flop: Consists of two conditioning inputs (J and K) and one clock input.

Karnaugh map: Truth table that facilitates reduction or simplification of a Boolean expression.

Latch: A logic circuit that can store an output.

Least significant bit (LSB): Digit with the lowest weighting in a binary number.

Logic gate: Gates signal transmissions according to the application, removal, or combination of input signals.

Look-ahead adder: "Looks ahead" to determine that all carries generated are available for addition.

Memory: A circuit that can store data.

Memory address register: Holds the data currently selected.

Memory cycle time: The minimum time between the reading of two words.

Memory data register: Passes all data written into and read from the memory.

Most significant bit (MSB): Digit with the highest weighting in a binary number.

Multiplexer: A data selector.

NAND gate: Combined NOT and AND functions in a binary circuit that has two or more inputs and one output.

N-bit comparator: Compares the magnitude of two numbers X and Y.

Negative logic: Has zero as the on state and one as the off state.

Noise margin: Is the limit in the tolerance spread and the degree of loading of a gate.

Non-volatile memory: Retains data even if the memory power is lost.

NOR gate: Combined NOT and OR functions in a binary circuit that has an output of logic 0 if any of the inputs is logic 1, and an output of logic 1 if all of the inputs are 0.

Octal numbering: Numbering system based on the powers of 8.

OR gate: Has a low output only when all of the inputs are low.

Parity bit: An extra bit added to the data block for data verification.

Parallel A/D converter: Input signal is compared to several reference signals by as many comparators. Output is low when the input signal is less than or equal to the reference signal; otherwise, it's high.

Parallel entry: When all the bits of data are transferred at the same time.

Positive logic: When one is the on state and zero is the off state.

Programmable read-only memory (PROM): A ROM that can be programmed by the user only once.

Programming: Preparing a permanent list of instructions for a computer to use in the solution of a problem.

Propagation delay: The time required for a signal to travel through a gate.

Random access memory (RAM): Can select data in memory in any order.

Reading: Obtaining data from memory.

Read-only memory (ROM): Used only for permanent data storage. Usually has no rewriting capability.

Refresh: Recharging memory capacitors.

Register: An assembly of flip-flops capable of storing binary data.

Resistor-transistor logic (RTL): Has a resistor as the input component, coupled to the base of an npn transistor. Produces the positive NOR or negative NAND gate functions.

Ring counter: Can store several bits of information. It accepts shift instructions that cause the information to shift one position at a time. The information recycles every *n* shift pulse, where *n* is the number of bits in the ring counter.

Ripple adder: Binary adding system in which the column of lowest order is added, then the resulting carry is added to the column of the next highest order, and so on for all columns.

RS flip-flop: Has two inputs, R and S. A 1 on the S input sets the flip-flop to the 1 or "on" state, and a 1 on the R input sets the flip-flop to the 0 or "off" state. It is assumed that a 1 will not appear on both inputs at once.

Serial entry: The transfer of data one bit at a time.

Sequential address: Linear data such as on magnetic tape, or cyclic data such as on a long shift register.

Settling time: Time required for a D/A converter output to arrive at and stay within one bit of its stable value.

Square waveform: An ac periodic waveform. Voltage alternates rapidly from a positive peak value to a negative peak value and vice versa, after a delay.

State: Condition at the output of a circuit that represents logic 0 or logic 1.

Static memory: Semiconductor memory in which the basic storage element can be set to either of two states, in which it will remain as long as the power stays on.

Successive approximation A/D converter: Input signal is compared with a D/A converter output one bit at a time.

Sum of products: General form of Boolean algebra that can be implemented through the use of electronic gate circuits.

Synchronous counter: Counter with a constant time interval. The sequence of operations is controlled by equally spaced clock signals or pulses.

Synchronous operation: Operates dependently on the clock pulse.

Truth table: Shows the relation of all output logic levels of a digital circuit to all possible combinations of input logic levels such that the circuit functions are characterized completely.

Volatile memory: Loses data when it loses power.

Wired logic: Positive logic used to reduce the number of gate leads required for a logic implementation.

Word-oriented: Also known as a linear select system. Two-dimensional system with a sense line (for reading) connected to the memory data register input and a write line connected to the memory data output.

Writing: Placing temporary data into memory.

XOR gate: Has a low output when all the inputs are the same.

INDEX

Symbols

0 level 4
1 level 4
1/10 decoder 60
1/2n decoder 60
1/n decoder 60
1/n^2 decoder 60
2421 code 45
8421 code 43, 141

A

access 93, 141
access time 93, 141
accumulator 87
address 93, 141
address bit 63
algebra of switching circuits 19
American Standard Code for Information Interchange (ASCII) 46, 141
analog 3, 101
analog circuit/signal 3, 4, 101, 141
analog comparator 103
analog vs. digital 3, *3-5*
analog-to-digital (A/D) converter 101, *101-108*, 103, 141
AND gate 7, 8, 141
anticipated carry adder 87
arithmetic circuits 53, 85, *85-92*
associativity 69
asynchronous 49
asynchronous counter 75, 141
asynchronous inputs/output 141
asynchronous jam transfer 53
asynchronous operation 50, 141
automobiles 110

B

basic arithmetic operations 42
basic-bit multiplier 91
binary 75
binary arithmetic 42, 141
binary code 41, 43
binary coded decimals (BCD) 44, 141
binary data 49
binary DOWN counter 75
binary number system 41, 43
binary UP counter 75
binary weighted ladder 101
binary weighted ladder D/A converter 103
biquinary code 46, 142
bit 73, 85
bit-oriented memory 94, 142
Boolean algebra 19, 24, 142
Boolean connective 19

C

calculus of propositions 19
carry 85, 87
character-to-character ripple adder 89
clock pulse period 76
clocked & gated flip-flop 30
CMOS circuits 16
CMOS inverter 16
codes 41, *41-48*
coincident signal select 94, 142
combinational logic circuit 37
commutativity 69
comparator 142
comparator circuit 69, *69-74*, 73
comparator outputs 103
complementary metal-oxide semiconductor logic (CMOS) 16
complementary outputs 63
complementing ring counter 54
computers 109, 110
control circuit 37, *37-40*, 142
control signals 41
controlled inverter 71
controlled shift register 53
converter 89
counter 53, 75, *75-84*, 77, 142
counter A/D converter 106, 107, 142
counter conversion 103
counter design 80
current 101
current sinking 11, 142
current sourcing 11, 142

cyclical code 45

D

D flip-flop 34, 142
D/A converter 104
data router 63
data selector 57, 63
decade counter 75
decade decoder 60
decimal number system 41
decode 59
decoder 57, *57-67*, 142
DeMorgan's theorem 24
demultiplexer 64, 142
digital 3
digital circuit 4, 142
digital code 41, 142
digital computers 43
digital data 93
digital signal 3, 101
digital system 101
digital-to-analog (D/A)
 converter 101, *101-108*,
 142
diminished complementing
 ring counter 55
diode transistor logic (DTL)
 13, 14, 142
direct clear & direct set inputs
 33, 34
distributivity 69
"don't care" state 80
DTL 13
DTL circuit 14
DTL gate 13
dual integration 106
dual slope A/D converter
 104, 106, 143
dual slope conversion 103
dual tree decoder 59

dumping 93, 143
dynamic memory 94, 143

E

ECL gate 15
edge-triggered flip-flop
 50, 75, 80
education 110
eight-line-to-one-line
 multiplexer 63
emitter coupled logic 15
enable input 63
enable signal 50
encoder 57, *57-67*, 143
EOC (end of conversion)
 signal 106
equality detector 73, 143
erasable and reprogrammable
 memory (EAROM) 94,
 143
error detection 46
event counters 75
excess-3 code 45, 143
exclusive-OR (XOR) circuit
 7, 8, 69, *69-74*, 143
external memory 93, 143

F

fan-in 11, 143
fan-out 13, 143
fast adder system 89
flip-flop *29-36*, 143
four-line-to-one-line
 multiplexer 63
four-variable Karnaugh map
 24
full-adder 85, 143

G

gate 11

Gray code 45, 46, 143
Gray-code-to-binary-code
 converter 71

H

half adder 85, 143
hamming code 46, 143
higher-order flip-flops 76

I

inequality detector 73, 143
input conditions 37
input devices 37
input signals 37
integrated circuits 73
integrated-circuit flip-flops 76
internal memory 93, 143
INTERRUPT 39
INTERRUPT input 39
inverter (NOT gate) 5, 7, 143

J

JAM operation 49
JK flip-flop 34, 80, 144
JK master-slave flip-flops 75
Johnson counter 54

K

Karnaugh map 21, 144
keyboard buffer register 57

L

latch 144
leakage resistance 94
least significant bit (LSB) 73,
 144
letters 41
linear select system 94, 146
liquid crystal display
 technology 110

logic circuits 41
logic families 11, *11-17*
logic function implementation *19-28*
logic gate 7, *7-10*, 144
look-ahead adder 87, 144
LSB first-serial relative magnitude detector 73

M

machine code 41
master-slave flip-flop 33, 75, 80
medium scale integration (MSI) 58
memory 93, *93-99*, 144
memory address register 94, 144
memory cycle time 93, 144
memory data register 94, 144
memory enable input (ME) 94
memory expansion 96
memory plane 94
microminiature digital electronic circuits 109
modules 60
most significant bit (MSB) 45, 144
MSB first-relative magnitude detectors 73
MSI decoder 60, 64
multilevel circuit 24
multiplexer 57, *57-67*, 63, 144
multiplexer applications 63
multiplier 91

N

n-bit comparator circuit 73, 144
NAND gate 8, 58, 144
NAND sequential control circuit 39, 40
negative logic 144
negative logic device 4
noise margin 11, 144
non-volatile memory 93, 144
NOR gate 8, 58, 144
NOR sequential control circuit 39
NOT gate 7
numbers 41

O

O/R 39
octal number system 42, 144
OFF state 4
ON state 4
one-step parallel entry 49
OR gate 7, 8, 144
output level 37
OVERRIDE condition 39, 40
OVERRIDE inputs 39

P

parallel A/D converter 103, 104, 145
parallel adder 87
parallel conversion 103
parallel entry 49, 145
parallel-carry connected flip-flops 80
parallel-carry counter 80
parallel-to-serial converter 63
parity 46
parity bit 145
parity check circuit 71
partial product 91
polynomial 41
positive logic 20, 145
positive logic circuit 5
positive logic device 4
preclearing 49
presetting 49
product function 20
product of sums 20
programmable read-only memory (PROM) 94, 145
programming 93, 145
propagation delay 13, 76, 145
propagation time 87
pseudosynchronous counter 77
pseudosynchronous operation 77

R

R-2R ladder 101
R-2R ladder D/A converter 103
random access memory (RAM) 94, 145
read-only memory (ROM) 93, 94, 145
reading 93, 145
rectangular decoder 58, 59
reference analog signals 103
refresh 145
register 49, *49-56*, 145
resistor-transistor logic (RTL) 13, 145
ring counter 54, 145
ripple 87
ripple adder 85, 87, 145
ripple counter 75, 76

ROM format 97
RS flip-flop 145
RTL gate 13

S

S/C 39
semiconductor memories 94
sequence of states 54
sequential address 94, 146
serial carry 79
serial data 73
serial entry 53, 146
serial magnitude detector 73
serial-carry counter 80
settling time 101, 146
Shannon, C.E. 19
shift counter 54
shift registers 53
signed-twos complement 89
sinusoidal waveform 3
sixteen-line-to-one-line
    multiplexers 63
square waveform 3, 146
START condition 39
START signal 39
state 53, 146
static memory 94, 146
STOP input 39
STOP signals 39
successive approximation A/D
    converter 104, 146
successive approximation
    conversion 103, 106
sum 85
sum function 20
sum of products 20, 146
switching theory 19, 20
synchronous 49
synchronous counter
    75, 77, 79, 146

synchronous flip-flops 87
synchronous jam entry 50
synchronous operation
    50, 146

T

television LCD screen 110
three-bit multiplier 91
timing 50, 75
timing diagram 50
toggle flip-flop 75
transistor-transistor logic
    (TTL) 14, 15
transmission errors 46
tree decoder 58
truth table 5, 146
TTL gates 14, 16
TTL logic gates 14
two-bit counter 63
two-level signal 3
two-state devices 93
two-step operation 50
two-step parallel entry 49
two-to-one multiplexer 63

U

unknown analog signal 103

V

vacuum tubes 109
Venn diagram 21
volatile memory 93, 146

W

weighted code 43
wired logic 20, 146
word size 96
word-oriented memory 94,
    146
writing 146

writing into memory 93

X

XOR gate 7, 8, 71, 14 *(see
    exclusive-OR gate)*

PROMPT® Publications is your best source for informative books in the technical field.

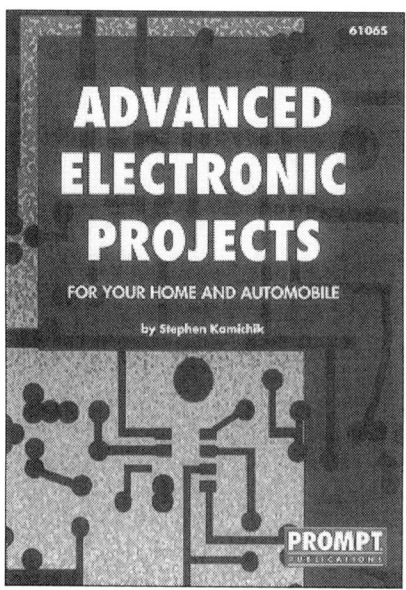

Advanced Electronic Projects for Your Home and Automobile
by Stephen Kamichik

You will gain valuable experience in the field of advanced electronics by learning how to build the interesting and useful projects featured in *Advanced Electronic Projects*.

$18.95

Paper/160 pp./6 x 9"/Illustrated
ISBN#: 0-7906-1065-5
Pub. Date 5/95

Semiconductor Essentials
by Stephen Kamichik

Gain hands-on knowledge of semiconductor diodes and transistors with help from the information in this book. *Semiconductor Essentials* is a first course in electronics at the technical and engineering levels. Each chapter is a lesson in electronics, with problems presented at the end of the chapter to test your understanding of the material presented.

$16.95

Paper/112 pp./6 x 9"/Illustrated
ISBN#: 0-7906-1071-X
Pub. Date 9/95

Call 1-800-428-7267 TODAY for the name of your nearest PROMPT® Publications distributor. Be sure to ask for your free PROMPT® catalog!

PROMPT® Publications is your best source for informative books in the technical field.

Speakers for Your Home and Automobile

How to Build a
Quality Audio System

by McComb, Evans & Evans

With easy-to-understand instructions and illustrated examples, this book shows how to construct quality home speaker systems and how to install automotive speakers.

$14.95

Paper/164 pp./6 x 9"/Illustrated
ISBN#: 0-7906-1025-6
Pub. Date 10/92

Sound Systems for Your Automobile

The How-To Guide for Audio System
Selection and Installation

by Alvis Evans and Eric Evans

Whether you're starting from scratch or upgrading, this book will show you how to plan your car stereo system, choose components and speakers, and install and interconnect them to achieve the best sound quality possible.

$16.95

Paper/124 pp./6 x 9"/Illustrated
ISBN#: 0-7906-1046-9
Pub. Date 1/94

**Call 1-800-428-7267 TODAY for the name of your nearest PROMPT® Publications distributor.
Be sure to ask for your free PROMPT® catalog!**

PROMPT® Publications is your best source for informative books in the technical field.

Advanced Speaker Designs for the Hobbyist and Technician

by Ray Alden

This book shows the electronics hobbyist and the experienced technician how to create high-quality speaker systems for the home, office, or auditorium.

$16.95

Paper/136 pp./6 x 9"/Illustrated
ISBN#: 0-7906-1070-1
Pub. Date 7/94

Security Systems for Your Home and Automobile

by Gordon McComb

You can save money by installing a security system yourself. This book tells you everything you need to know to select and install a security system with a minimum of tools.

$16.95

Paper/130 pp./6 x 9"/Illustrated
ISBN#: 0-7906-1054-X
Pub. Date 7/94

Call 1-800-428-7267 TODAY for the name of your nearest PROMPT® Publications distributor. Be sure to ask for your free PROMPT® catalog!

PROMPT® Publications is your best source for informative books in the technical field.

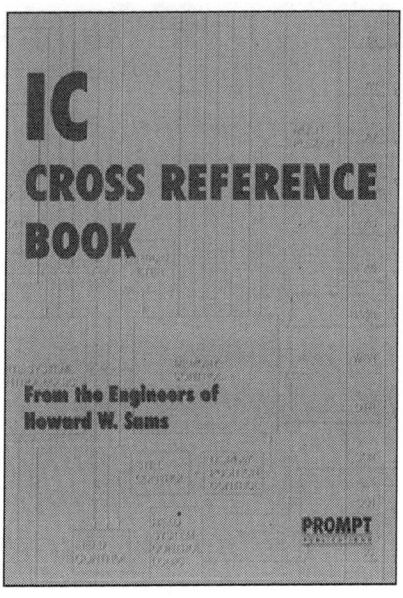

Semiconductor Cross Reference Book
Revised Edition
by Howard W. Sams & Company

From the makers of PHOTOFACT® service documentation, the *Semiconductor Cross Reference Book* is the most comprehensive guide to semiconductor replacement data. The volume contains over 475,000 part numbers, type numbers, and other identifying numbers.

$24.95
Paper/668 pp./8-1/2 x 11"
ISBN#: 0-7906-1050-7
Pub. Date 6/94

IC Cross Reference Book
by Howard W. Sams & Company

The engineering staff of Howard W. Sams & Company has assembled the *IC Cross Reference Book* to help you find replacements or substitutions for more than 35,000 ICs or modules. It has been compiled from manufacturers' data and from the analysis of consumer electronics devices for PHOTOFACT® service data, which has been relied upon since 1946 by service technicians worldwide.

.$19.95
Paper/168 pp./8-1/2 x 11"
ISBN#: 0-7906-1049-3
Pub. Date 5/94

Call 1-800-428-7267 TODAY for the name of your nearest PROMPT® Publications distributor.
Be sure to ask for your free PROMPT® catalog!

PROMPT® Publications is your best source for informative books in the technical field.

Surface-Mount Technology for PC Boards

by James K. Hollomon, Jr.
Manufacturers, managers, engineers, and others who work with printed-circuit boards will find a wealth of information about surface-mount technology (SMT) and fine-pitch technology (FPT) in this book.
$26.95
Paper/309 pp./7 x 10"/Illustrated
ISBN#: 0-7906-1060-4
Pub. Date 7/95

Electronic Control Projects for the Hobbyist and Technician

by H. C. Smith and C. B. Foster
Would you like to know how and why an electronic circuit works, and then apply that knowledge to building practical and dependable projects that solve real, everyday problems? Each project in *Electronic Control Projects* involves the reader in the actual synthesis of a circuit.
$16.95
Paper/168 pp./6 x 9"/Illustrated
ISBN#: 0-7906-1044-2
Pub. Date 11/93

Call 1-800-428-7267 TODAY for the name of your nearest PROMPT® Publications distributor. Be sure to ask for your free PROMPT® catalog!

PROMPT® Publications is your best source for informative books in the technical field.

Schematic Diagrams
The Basics of Interpretation and Use

by J. Richard Johnson

Step-by-step, *Schematic Diagrams* shows you how to recognize schematic symbols and their uses and functions in diagrams. You will also learn how to interpret diagrams so you can design, maintain, and repair electronics equipment.

$16.95
Paper/208 pp./6 x 9"/Illustrated
ISBN#: 0-7906-1059-0
Pub. Date 9/94

Industrial Electronics for Technicians

by J. A. S. Wilson and J. Risse

Industrial Electronics for Technicians provides an effective overview of the topics covered in the Industrial Electronics CET test, and is also a valuable reference on industrial electronics in general.

$16.95
Paper/352 pp./6 x 9"/Illustrated
ISBN#: 0-7906-1058-2
Pub. Date 8/94

Call 1-800-428-7267 TODAY for the name of your nearest PROMPT® Publications distributor.
Be sure to ask for your free PROMPT® catalog!

PROMPT® Publications is your best source for informative books in the technical field.

Tube Substitution Handbook
Complete Guide to Replacements
for Vacuum Tubes & Picture Tubes
by W. Smith and B. Buchanan

The most accurate, up-to-date guide available, the *Tube Substitution Handbook* is useful to antique radio buffs, old car enthusiasts, ham operators, and collectors of vintage ham radio equipment.

$16.95
Paper/ 149 pp./6 x 9"/Illustrated
ISBN#: 0-7906-1036-1
Pub. Date 12/92

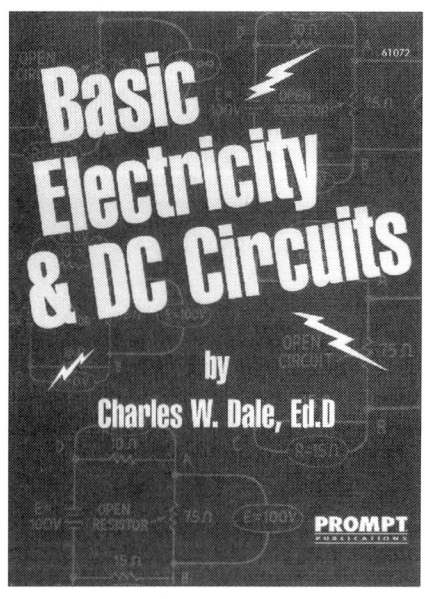

Basic Electricity & DC Circuits
by Charles W. Dale, Ed.D

Now you can learn the basic concepts and fundamentals behind electricity and how it is used and controlled. *Basic Electricity and DC Circuits* shows you how to predict and control the behavior of complex DC circuits.

$34.95
Paper/928 pp./6 x 9"/Illustrated
ISBN#: 0-7906-1072-8
Pub. Date 8/95

Call 1-800-428-7267 TODAY for the name of your nearest PROMPT® Publications distributor. Be sure to ask for your free PROMPT® catalog!

PROMPT® Publications is your best source for informative books in the technical field.

Basic Electricity
Revised Edition,
Complete Course
by Van Valkenburgh, Nooger & Neville, Inc.
From a simplified explanation of the electron to AC/DC machinery, alternators, and other advanced topics, *Basic Electricity* is the complete course for mastering the fundamentals of electricity.
$19.95
Paper/736 pp./6 x 9"/Illustrated
ISBN#: 0-7906-1041-8
Pub. Date 2/93

Basic Solid-State Electronics
Revised Edition,
Complete Course
by Van Valkenburgh, Nooger & Neville, Inc.
A continuation of the instruction provided in *Basic Electricity*, this book provides the reader with a progressive understanding of the elements that form various electronic systems.
$24.95
Paper/944 pp./6 x 9"/Illustrated
ISBN#: 0-7906-1042-6
Pub. Date 2/93

Call 1-800-428-7267 TODAY for the name of your nearest PROMPT® Publications distributor.
Be sure to ask for your free PROMPT® catalog!